Understanding ATM

Stan Schatt

McGraw-Hill

New York San Francisco Washington, D.C. Auckland Bogotá
Caracas Lisbon London Madrid Mexico City Milan
Montreal New Delhi San Juan Singapore
Sydney Tokyo Toronto

McGraw-Hill

A Division of The **McGraw·Hill** *Companies*

Library of Congress Cataloging-in-Publication Data

Schatt, Stan.
    Understanding ATM / by Stan Schatt
        p.   cm.
    Includes index.
    ISBN 0-07-057679-3 (pbk.)
    1. Asynchronous transfer mode.   2. Local area networks (Computer
networks) 3. Computer network architectures. I. Title.
TK5105.35.S48   1996
004.6'6—dc20   96-33776
                                          CIP

pbk 1 2 3 4 5 6 7 8 9 FGR/FGR 9 0 0 9 8 7 6

ISBN 0-07-057679-3

The acquisitions editor of this book was Jennifer Holt DiGiovanna, the executive editor was Robert E. Ostrander, the book editor was Kellie Hagan, and the director of production was Katherine G. Brown. This book was set in ITC Century Light. It was composed in Blue Ridge Summit, Pa.

Printed and bound by Quebecor, Fairfield, Pa.

McGraw-Hill books are available at special quantity discounts to use as premiums and sales promotions, or for use in corporate training programs. For more information, please write to the Director of Special Sales, McGraw-Hill, 11 West 19th Street, New York, NY 10011. Or contact your local bookstore.

MH96
0576793

This book is dedicated to my wife, Jane, who teaches the very lucky kindergarten children at Capri School all the important things they need to know to become happy and successful adults. Janie, you've enriched my life as well as theirs.

Contents

Chapter 2. ATM Basics   41

Chapter 3. Moving Toward Virtual Networks and ATM   67

Part 2   Vendor Network Architectures and ATM Plans

Introduction

Several years ago I wrote one of the first books on local area networks. *Understanding Local Area Networks* was published in 1987, a time when most people still hadn't even heard the term. What excited me about LANs, as early as 1985, was the potential I saw for the technology. Two million LAN nodes and five editions of the book later, the public seems to have vindicated my early faith in this technology.

I have the same feeling about asynchronous transfer mode (ATM). The vast majority of people still associate the term with banking (automatic teller machines) and not with high-speed communications, yet the impact of ATM could be just as dramatic as that of local area networks on the way companies conduct their business.

One of the dangers of being at the "bleeding edge" of a new technology is the possibility of falling in love with the technology and ignoring practical considerations. ATM has progressed far faster than even its most optimistic supporters would have dreamed. After watching a major high-bandwidth technology's window of opportunity close while an industry committee debated the merits of FDDI, vendors have worked together remarkably smoothly on the ATM Forum to produce the first round of specifications. Given the cost of ATM equipment, very few network managers want to buy proprietary technology and wind up located into a single vendor's product line. ATM is real and it's here today.

The purpose of this book is to provide an overview of ATM technology and a snapshot of today's ATM product directions. While IBM or 3Com might offer a new switch or a new hub after this book comes out, the book's focus is on product families and vendor product architecture. IBM's products will change, but they'll be consistent with the company's switched virtual networking architecture. Similarly, Bay Networks' products appear almost monthly, but they're all consistent with the company's BaySis architecture.

By examining the overall directions companies are taking, you can see gaps in a company's architecture or in its product offerings. Because the industry is so new, it's helpful to compare and contrast product features to see which are proprietary and which conform to new ATM Forum standards.

This book is divided into two sections. The first is concerned with key concepts, and is written for readers with no prior background. The problem with ATM is that its implementation requires a knowledge of local area networks, virtual networks, and wide area network technology, so most readers will have some gaps in their knowledge that these first chapters can fill in.

Chapter 1 looks at local area networks in terms of their key component building blocks. It covers a wide range of LAN topologies, such as 100-Mbps Ethernet and Token Ring, and the role of network interface cards, servers, and network operating systems. It even looks at LAN switches and offers some information on how to evaluate them. Chapter 2 turns toward ATM. It serves as a basic grounding in the key principles associated with this technology, including its protocols, switches, cell structure, and network management. One section explains why LAN emulation over ATM is so important and then describes how the process works. Finally, chapter 3 looks at the concept of virtual networks. This is a way to construct logical rather than physical networks and is a key element in a switched-based network environment such as ATM. This chapter is probably the most detailed available explanation on this complex topic and should allow you to understand how various vendors' virtual networking products differ.

Many early ATM buyers are looking toward a time in the future when they can have seamless local-area-to-wide-area-network ATM communications. The problem is that many network managers lack a thorough background in telecommunications. This field has its own obscure abbreviations and terminology and can be very confusing. So this chapter presents the basics of services such as frame relay, ISDN, SMDS, and ATM, along with explanations on such key wide area components as packet switching, T-1 and T-3 lines, and even fractional T-1 lines.

The second half of this book looks at specific vendors' network architecture, ATM products and plans, and their overall vision of a switched ATM enterprise environment. Several vendors receive an entire chapter, including FORE Systems, IBM, Digital Equipment Corporation, and Cisco Systems. A separate chapter looks at the leading hub vendors' vision of ATM and looks at 3Com, Bay Networks, Cabletron Systems, and UB networks. Chapter 10 examines the state of wide area network ATM technology and products today. I'll describe some of the core switches designed for carriers' central offices, as well as proliferating edge switches and the unique role they play. Chapter 11 is devoted to desktop ATM and looks at two key players in this market: Madge Networks and First Virtual Corporation. Included is a description of how multimedia could become the "killer" application for desktop ATM.

The last two chapters in this book are designed to help you evaluate ATM products and find additional information on the topic. Chapter 12 consists of several checklists you can use to compare and contrast products. Chapter 13 explains how to find books, articles, seminars, etc. on the topic. There's also a glossary that provides definitions of key terms used in the book.

ATM is an exciting technology because it can change the way companies conduct business. Will there be more multimedia desktop applications, more high-speed LAN-to-WAN communications, and high-speed LAN backbones that can support video conferencing at sites? The odds look very good at present.

I hope this book provides you with good, usable information to help you understand the changes taking place in the network industry and the way companies will migrate from shared media networks to switched networks.

Stan Schatt
Carlsbad, California

Key ATM Concepts

Today's Corporate
Network Environment

While I wrote this book to help you understand asynchronous transfer mode (ATM), this new technology will coexist with currently installed computer equipment. In this chapter, I'll describe the current network environment, focusing on how local area networks are designed as well as how they operate. You'll learn how each key component of a local area network operates, why traffic congestion occurs, and some of the many technologies that offer higher bandwidth for overtaxed networks. You'll see how Ethernet and Token Ring switches operate and why they can often solve network traffic problems. You'll also examine the two versions of fast Ethernet currently available and compare the advantages and disadvantages of each.

This chapter will provide you with a thorough enough understanding of local area network technology to understand the ATM and local area network interoperability issues that are the subject of chapter 3. Remember, I'm assuming that you're just a beginner when it comes to networking, so the pace will be relaxed. Even experienced network managers, however, should find some of the information new and interesting.

Technology alters the office environment so quickly that it's easy to lose sight of the explosive changes in the way people work that have taken place over the past decade. In 1995, the Smith-Corona company declared bankruptcy. The leader of a typewriter industry that had penetrated virtually every office by 1980 saw its sales evaporate as customers shifted to stand-alone personal computers (PCs) with word-processing programs and printers. By the mid-1980s, many companies had begun linking together PCs to form networks in order to share equipment and programs.

The 1990s have seen a proliferation of networks found at a single location (local area networks) as well as the growth of wide area networks that link corporate networks at several locations regionally, nationally, or even globally. For reasons that will be explained later in this chapter, many companies are now planning to adopt a new technology known as asynchronous transfer mode (ATM). In order to understand why ATM is becoming so desirable, you must first understand local area network technology.

What Is a Local Area Network?

Industry market research leader Computer Intelligence InfoCorp believes that in 1995 roughly 6 out of every 10 personal computers (PCs) found in U.S. businesses were networked. A *local area network (LAN)* consists of a communication network used by a single organization over a limited distance that permits users to share information and resources. This limited distance is usually less than one mile. There are two main reasons why LANs have grown so rapidly. Companies find they save considerable money by being able to spread the cost of expensive hardware such as high-speed laser printers over dozens or even hundreds of users. It's much easier to justify the considerable cost of a color laser printer by pointing out that everyone in the company will have access to it. Similarly, LAN software licenses often are more economical than purchasing stand-alone versions of a program for each user.

An even more significant reason for the growth of LANs has been an increase in worker productivity due to the ability to share common software. That means, for example, that all salespeople can access the same customer database and add information, or that all financial people have access to the same accounting information.

While people can share information via mainframes or minicomputers, the advantage of using a LAN is that users can leverage the processing power of their personal computers (PCs). For example, a LAN user requests information from a PC that serves as a file server. It serves up the required files but doesn't need to perform all the processing a mainframe or minicomputer is required to do when providing information to a dumb terminal. The term *distributed processing* is often used to describe a LAN environment in which users' microcomputers perform their own processing rather than depending on the centralized processing of larger computers.

While the major focus of this book is on ATM, this technology won't exist in a vacuum. Chapter 3 describes how companies are likely to create networks in which older LAN technology (legacy LANs) will coexist with ATM products. In order to understand the technical, financial, and even political issues involved in creating such environments, it's necessary to understand some basic LAN concepts. That's the purpose of this chapter.

Figure 1.1 shows a typical local area network. The individual PCs each have a network interface card (NIC) installed in one of their motherboard slots. This circuit card has enough intelligence to be able to transmit and receive messages over the LAN. Cabling (the media) connects the NIC to the LAN infrastructure. The way the network is structured is known as its *topology*. The ring-like topology shown in Figure 1.1 is known as Token Ring architecture. A file server provides and stores information as well as coordinating the use of all hardware and software resources. It uses network operating system software (NOS) to help it perform these tasks. Now that you have the big picture of a LAN in action, let's spend some time looking at the various types of available LANs as well as the basic building blocks of a LAN.

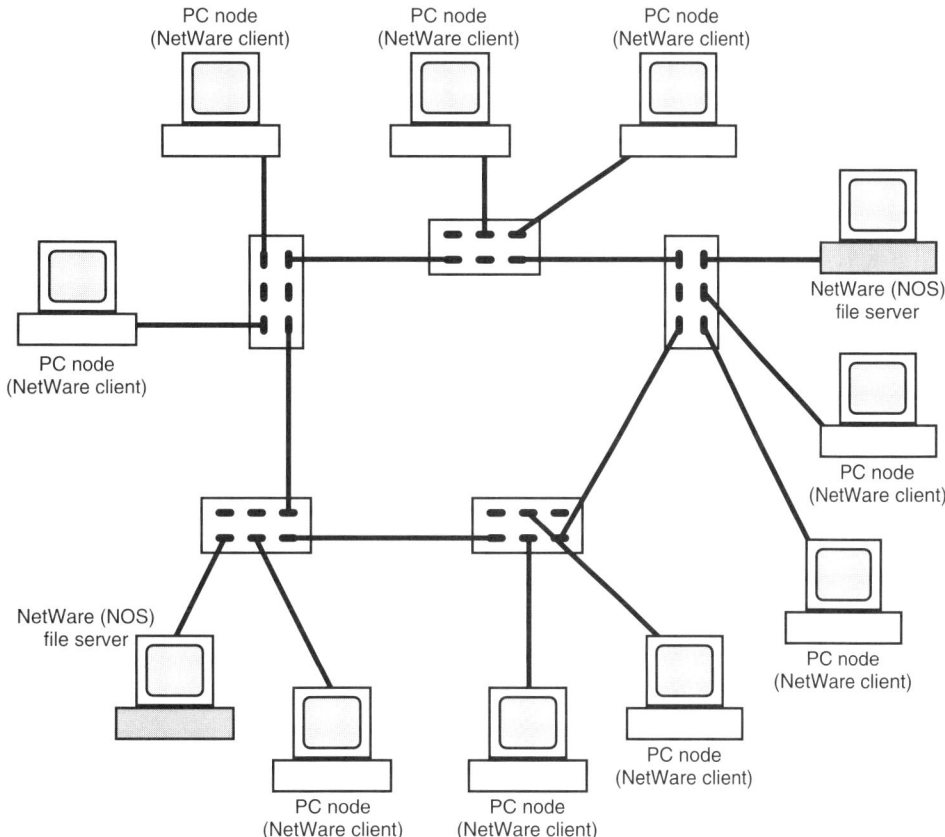

Figure 1.1   A typical local area network and its basic building blocks; this one is a Token Ring network with concentrators attached via cabling to network interface cards in PCs.

LANs Come in Different Shapes and Sizes

LANs have evolved from small workgroup-sized networks coexisting with IBM mainframe computers that perform most of the corporate computing to corporate-wide enterprise LANs that can include thousands of users. In later chapters, I'll describe how ATM can coexist with present LAN technology, a category that ATM vendors refer to as *legacy LANs*. In order to understand what this term really means, it helps to examine the different forms LANs can take today and how they differ. We'll start with the earliest and still most popular LAN topology (network architecture): Ethernet. Before you look at network topology in general and Ethernet specifically, however, it's necessary to understand how different types of programs that exist on a network pass information to each other. That's really another way of describing the open systems interconnect (OSI) model. This is a complex subject, but I'll try to make it as painless as possible.

The OSI Model

In order for computers to communicate with each other and exchange information, a number of complex issues have to be resolved between them. Imagine that two countries have decided to exchange information. Diplomats might have to negotiate issues such as what language (syntax) will be used, what alphabet will be used (how the language will be displayed), where the communications will be sent (the addresses to be used), and the priority these communications can take (is a presidential message guaranteed the highest priority?). Other questions might involve how the physical link will be established (where will the telephone lines be installed?) and the diplomatic procedure required to establish communications (will a password be required?). All these issues (modified, of course, to reflect the world of electronic communications) need to be resolved every time a message is sent over a local area network.

The International Organization for Standardization (ISO) in conjunction with the Consultative Committee on International Telegraphy and Telephony (CCITT) developed a model to facilitate communications between computer networks. As shown in Figure 1.2, the open systems interconnect (OSI) model specifies a series of software layers, each with their own rules and procedures (protocols) for accomplishing specific tasks. Since the OSI model is the basis for network communications regardless of whether you're talking about Ethernet or ATM, I'll spend a few moments examining the role of each of these key protocol layers. Later in this chapter I'll return to this topic and show you how this OSI model is the basis of what happens inside a network interface card.

To understand how the OSI model works, it's necessary to understand the concept of layered protocols. Imagine a typical company with a hierarchical (layered) organization. Each department in the organization chart receives orders from the department above it. It uses a corporate manual of procedures and policies (protocols)

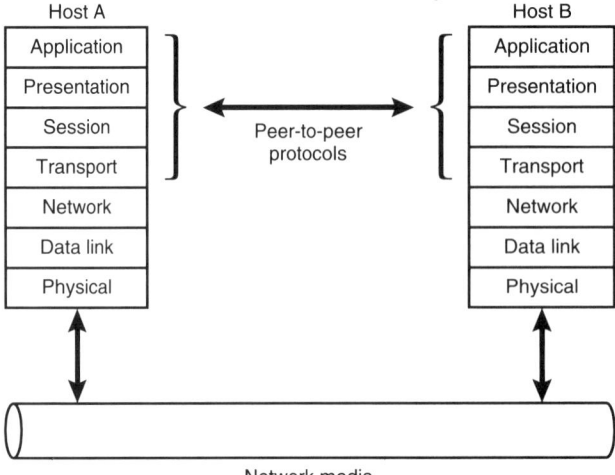

Figure 1.2 The OSI model.

to determine how to handle these tasks. If an order is received that requires the department to provide information in a certain format, it follows those orders precisely. In turn, it hands down orders to the department under it with precise directions as to how it wants its information handled.

This chain of command works very well. To extend the analogy a bit further, imagine that each department requires information presented in a certain format. Each time a department receives data (in its own required format), it processes and then reformats the data to meet the demands of the department above it. This repackaging of data can mean adding instructions. After the data has gone down to the mail room (the lowest department), it might have several different instructions attached to it, associated with the various departments whose input was required. These instructions remain with the corresponding data because they'll be needed when the data arrives at a different company. There, it will move up the various departmental levels. Each department will follow the attached directions to ensure that the data can be understood and presented in the appropriate format to the appropriate department. Now, let's take a close look at the various OSI layers that corresponds to the departments in the example.

The application layer

The application layer is designed to handle the requests made by an application running on a network. Let's assume that an electronic mail (e-mail) user needs to send a message to another network user. The application layer has several protocols associated with handling the unique requirements of electronic messaging. It also handles such functions as file transfer, terminal emulation, electronic data interchange, and transaction processing.

The presentation layer

This layer concerns itself with how information is physically displayed or presented. Numbers, for example, might require a floating-point format or, conversely, might need to be displayed in currency format. Certain database fields might require both characters, numbers, and even graphics. If encryption must take place to protect the data from being intercepted and read by a network intruder, then this is the layer where encryption takes place.

The session layer

Think of the session layer as an administrative assistant with a good eye for detail who is responsible for arranging all the details for an important upcoming meeting between two executives. This layer is concerned about such basic issues as the mode of transmission and synchronizing (timing) points. In other words, will transmission take place in half-duplex (with each side alternating the sending and receiving of information) or full duplex (with the sending and receiving of information taking place simultaneously)?

Synchronizing points represent points during the "conversation" that the session layer checks to ensure that actual communication is taking place. If you've ever observed two Japanese businesspeople carrying on a conversation, you'll certainly remember that both were nodding their heads and saying "hi." This doesn't mean they agreed with each other; it simply means that they were indicating that they heard and understood what the other was saying, since "hi" means "yes" in Japanese.

Another concern of the session layer is how to reestablish communications if they're disrupted during a session. A synchronization point between pages of a text, for example, would be a logical approach since reestablishing a session wouldn't require beginning again and repeating all text already received correctly.

This layer is also responsible for the details associated with an orderly (known as *graceful*) release of the connection when a session ends. There are also procedures for what's known as an *abrupt* release of a connection, a situation where one side ends communications and refuses to receive additional data from that moment on.

The session layer doesn't automatically have to accept every connection. It might issue a refusal order if, for example, it determines that it would result in too much network congestion or because the procedure requested is unavailable for use.

The transport layer

The transport layer is of great importance to computer network users because it provides the quality of service required by the network layer. To clarify the transport layer's responsibilities, think of it as a collection of special services available for additional cost at your local post office. In other words, for an additional fee you can receive a return receipt indicating that a letter was delivered to the person you specified. Similarly, you might require expedited service, such as next-day delivery. While the U.S. Postal Service collects fees for these high-quality added services, the price paid by a computer user running OSI-compatible hardware and software are the additional bits required to provide information on the status of these possible services.

Three different types of network service are provided by the transport layer. Its type A service provides network connections with acceptable residual error rates and acceptable rates of signaled failures. Type B offers an acceptable residual error rate and an unacceptable rate of signaled failures, and type C provides network connections with residual error rates that aren't acceptable to the session layer.

Why have classes of service that provide unacceptable error rates? Many network connections use additional protocols that provide the required error detection and recovery, and the required overhead for this service under the transport layer isn't necessary.

On the other hand, the transport layer does allow programmers to write programs for the application layer for a wide variety of networks without concerning themselves with whether or not transmission on these networks is reliable. In fact, some people distinguish the top three layers of the OSI model as *transport layer users* and the bottom four layers as *transport layer providers*. The five classes of transport protocol service are listed in Table 1.1.

Class 0, known as Telex, is the simplest quality of service. It assumes that the network layer (below the transport layer) provides flow control, and it releases its connection when the network layer release its connection.

**TABLE 1.1 Classes of Transport Protocol
Service**

Class	Title	Type
0	Simple	A
1	Basic error recovery	B
2	Multiplexing	A
3	Error recovery and multiplexing	B
4	Error detection and recovery	C

Class 1 service was developed by the CCITT for its X.25 packet-switched network standard. It does provide expedited data transfers, but it still relies on the network layer for flow control.

Class 2 represents an enhanced class 0. The basic assumption made here is still that the network is highly reliable. The quality of service offered includes the ability to multiplex multiple transport connections from a single network connection. Class 2 ensures that multiplexed packets of data arriving out of order can be reassembled properly.

Class 3 provides the services offered by both 1 and 2, as well as the ability to re-synchronize so a connection can be reestablished if an error is detected.

Finally, Class 4 assumes that the network layer service is inherently unreliable. It offers its own error detection and recovery procedures.

The network layer

The network layer is where network routing takes place. I'll discuss routers later in this chapter, but routing refers to figuring out the best way to send data from one computer to another computer. As such, it's the key to understanding how gateways to IBM mainframe and other computer system functions. While upper-level OSI protocols request that a packet be transmitted from one computer system to another, the network layer concerns itself with the actual mechanics of the journey.

The network layer performs a number of key services for the transport layer immediately above it on the OSI model. It notifies the transport layer when it detects unrecoverable errors, and helps this layer maintain quality of service and avoid network congestion by stopping the transfer of packets whenever necessary.

Because physical connections can change from time to time when two networks are communicating, the network layer ensures that packets arriving out of sequence are reassembled correctly. In effect, it uses routing tables, which helps it determine which path a particular packet should take. Often a message composed of multiple packets will take different pathways. The network layer provides essential "shipping" information for these packets, such as the total number of packets comprising a message and the sequence number of each packet.

One very unfortunate complication in network communications is that different networks have different sized address fields, different sized packets, and even different time intervals during which they permit a packet to circulate before they *time out*, a condition where the packet is considered lost and a duplicate packet is requested. The network layer must provide enough control information within the packets to address these issues and ensure successful delivery and reassembly.

As mentioned earlier, there's a good deal of duplication between the functions of the transport layer and the network layer, particularly in flow control and error checking. The primary reason for this duplication is that there are two different types of connections, connection-oriented and connectionless, and they make far different assumptions regarding network reliability.

A *connection-oriented network* functions very much like the telephone system. Once a connection is established, communication continues, point to point, without the two parties feeling compelled to conclude each statement with their name, the name of the party they called, and that party's address. It's assumed that the communication is reliable and that the other party receives the message as it's sent.

Given the assumption of a reliable, connection-oriented network, the destination address is required only when the connection is established; individual packets don't have to carry this address in a separate field. The network layer assumes responsibility for error checking, as well as flow control, in a connection-oriented network. It also concerns itself with the actual sequencing of the packets.

A *connectionless network*, on the other hand, relies more on the transport layer for error checking, flow control destination addresses are required on each packet, and packet sequencing is not guaranteed. The basic assumption of this service is that speed is paramount and end users should have their own error-checking and flow-control software rather than rely on a standard built into the OSI model.

As is usually the case whenever committee members argue a complex issue, a compromise was reached that really didn't totally satisfy anyone. The compromise is that the capabilities for both connection-oriented and connectionless service are both built into the OSI model's network and transport layers. End users can select appropriate default values for these two layers' control fields and use the approach they prefer. The negative side to this compromise, however, is the significant amount of redundancy built into the two layers, which means a significant amount of bit overhead. Information in this OSI format being sent over long-distance lines translates into increased costs since the transmission takes longer.

The data link layer

Think of the data link layer as the warehouse and receiving/shipping dock of a large manufacturing company. It must take the packets it receives from the network layer and prepare them for transmission (shipping) by placing them in the appropriately sized packages (frames). When information is flowing through the layers of the OSI model, the data link layer must be able to take the raw bits coming from the physical layer and make sense of them. It must establish when a transmission block starts and where it ends, as well as detect whether there are transmission errors. If it does detect errors, this layer is responsible for initiating action to recover lost, garbled, and even duplicate data.

Several data links can exist simultaneously and function independently between computer systems. The data link layer is also responsible for ensuring that these transmissions don't overlap and that data doesn't become garbled. The data link layer initializes a link with its corresponding layer on a computer with which it will communicate. It has to ensure that both machines' clocks are synchronized and that they both use the same encoding and decoding schemes.

Since flow control and error checking are also the responsibility of the data link layer, it monitors the frames it receives and maintains statistical records. When a user has completed transferring information, the data link layer assumes responsibility for determining that all data was received before terminating the link. This layer also contains a number of protocols defined by the Institute of Electronic and Electrical Engineers (IEEE) 802 Committee that develops industry-wide specifications for local area networks.

The physical layer

The physical layer of the OSI model is the least controversial since it contains international hardware standards that have become commonplace. In fact, one real issue concerning this layer is how the ISO will handle emerging new hardware standards. Because the ISO and IEEE subcommittees have worked together for several years, it isn't surprising that many LAN standards use the definitions provided at the physical layer by the OSI model.

The OSI model's physical layer defines such key network components as the type of coaxial cabling or unshielded twisted-pair wire required to achieve a 10-Mbps transmission speed. This layer also specifies the type of encoding scheme a computer uses to represent binary values for transmission over a communication channel. By encoding scheme, I mean the type of electrical signals used to represent a binary 1 or a binary 0.

So the OSI model's physical layer assumes responsibility for the type of physical media, transmission, encoding method, and data rate associated with different types of local area networks. It's also responsible for establishing the physical connection between two communications devices, generating the actual signal, and then making sure the two devices are synchronized. The timing of the two units' clocks must be the same so transmitted information can be decoded and understood.

Network Topology

A network's *topology* or architecture/structure determines its ultimate effectiveness because inherent in the topology are a number of strengths and weaknesses. Imagine that a child is given a box of snap-together blocks and asked to build a bridge connecting two toy cities. Clearly, there are a number of ways to connect the blocks and a number of different shapes the bridge can assume and still carry out its function as a bridge. Some bridges might require less blocks and thus be far cheaper to construct. Other bridges would be so narrow that they would restrict the flow of toy cars to a single lane and would thus cause traffic congestion. Still other bridges would contain multiple car lanes and permit fast travel in both directions, yet would require more blocks that would add to the construction costs. Let's see how this analogy relates to network topology.

Ethernet

History records that Hitler and Mussolini once met to determine how the world would be carved up after their victory over the Allies. With far better success, Digital Equipment Corporation (DEC), Xerox, and Intel worked closely in the late 1970s and early 1980s to develop a local area network architecture that would capture the entire worldwide network market. With well over 60 percent of all LANs today, the three corporate giants have come close to realizing their goal. The plan for Ethernet was very simple. Xerox would provide its expertise in network operating system software, Intel would provide computer chip technology, and Digital would provide the computers.

Ethernet first appeared in 1980. When the industry standards group, the Institute of Electronic and Electrical Engineers's 802 Committee, met to set the specifications for a bus type network, Ethernet was already a *de facto* standard. The IEEE 802.3 set of specifications introduced some minor modifications, but left the networking technology pretty much intact. Let's take a closer look at what Ethernet is and how it works.

Figure 1.3 illustrates a typical Ethernet network from the late 1980s. Note that, while the first Ethernet LANs used very bulky, thick coaxial cabling, the vast majority of Ethernet LANs by 1990 were using the same thin coaxial cabling attached to your cable television box. A computer or *node* on such an Ethernet network contains an Ethernet network interface card that's responsible for handling the management, transmission, and reception of data—including the encoding and decoding of electronic signals. Transceivers on the NIC generate the electrical signals to the coaxial cabling and maintain their quality. They also receive network signals and detect any errors in transmission.

Ethernet is a bus network because it's laid out like as a data highway with straight sections known as *segments*. You can link segments together and use repeaters to extend a network's size. Feeder roads (or cables) connect individual computers to these network segments. There are a couple of important points about this kind of topology. Because all data broadcasts go down this central data highway, a break in the cabling will disrupt the entire network. Also, because of the dangers of possible electrical interference, there are rules for how close computers can be to each other and how many of them can exist on a single segment.

Accessing an Ethernet LAN. Ethernet has a bandwidth of 10 Mbps, which means that an Ethernet highway can move traffic at the rate of ten million bits per sec-

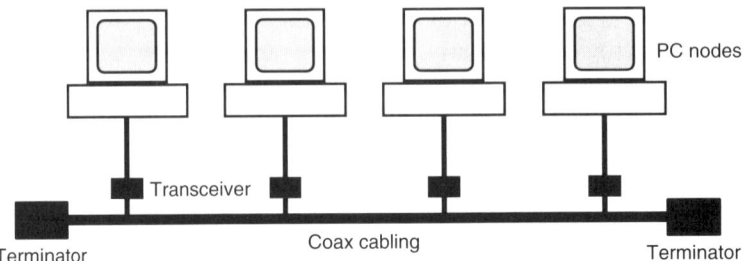

Figure 1.3 A typical Ethernet network with coaxial cabling.

Preamble	Destination address	Source address	Protocol type	**Data**	Frame check sequence

Figure 1.4 An Ethernet packet containing both data and control information.

ond. Of course, that's under absolutely ideal conditions. In reality, much of the data needs to be retransmitted because of data collisions on the network. The dark secret that plagues Ethernet is that it's a *contention* network topology and, as such, requires users to contend or compete for use of the network. Only one user can transmit data at a time. If two users transmit at the same time, there will be a data collision. Designed for networks at a time when bursty traffic was the rule, Ethernet works best when users transmit short bursts of data such as database inquiries ("find me the last purchase for Smith MegaWidgets"). In such an environment, there's a minimum of collisions because these transmissions reach their destinations so quickly.

Think of Ethernet as a very narrow one-lane highway running through town. Imagine that each resident who wants to drive to a store or neighbor's house must first place an ear down on the street and listen for sounds indicating that another car is approaching. If no sound is heard, then the resident would run and jump in the family car, back out of the driveway, and drive as quickly as possible toward the store or neighbor's house.

The only problem with this approach is that there's a brief time interval between when the resident listens for approaching traffic and when he actually backs out into the street and begins driving—during which another car could head down the street. When data collides, there's no messy crash requiring tow trucks or ambulances. A signal is generated over the LAN indicating that a collision has taken place, and Ethernet requires all network users to wait a predetermined amount of time before retransmitting data. Each network interface card waits a different length of time to decrease the likelihood of another collision when they begin transmitting data once again.

This particular method of accessing a LAN is known as *carrier sense multiple access/collision detection (CSMA/CD)*. Carrier sense refers to the fact that NICs listen to "sense" a signal, indicating that another user is transmitting data on the LAN. Multiple access refers to the fact that more than one user accesses the LAN, and collision detection refers to the way in which collisions are detected and handled.

Figure 1.4 illustrates what an Ethernet packet (actually the IEEE 802.3 close cousin of an Ethernet packet) looks like. Notice that control information is required so the data can be unpacked, successfully translated, and read on receipt. A data collision scrambles or destroys some of the bytes of information. Table 1.2 describes the contents of each field in this packet.

Table 1.3 summarizes some major advantages and disadvantages of Ethernet topology. *Fast* and *inexpensive* are two words that warm the hearts of most MIS and network managers. Traditional Ethernet (and its close cousin the IEEE 802.3 version) have been around so long that all LAN applications and network operating systems work with it.

TABLE 1.2 Contents of an Ethernet Packet

Field	Function
Preamble	A bit pattern that indicates an Ethernet packet
Destination address	The address of the destination node to receive this packet
Source address	The address of the node sending this packet
Protocol type	The type of protocol found in this packet
Data	The data contained in this packet
Frame check sequence	The bit pattern used to determine if data has been destroyed or damaged

TABLE 1.3 Advantages and Disadvantages of Ethernet Topology

Advantages	Disadvantages
Fast (10 Mbps)	Lack of guaranteed time for transmission
Inexpensive	Not efficient for heavy traffic
Good for bursty traffic	

The contention nature of Ethernet reflects the networking world at the time it was first developed, and contention networks function most efficiently with bursty traffic. This means that the short, rapid type of inquiries associated with requests for information from a database can be handled with little trouble or data collisions by Ethernet.

Where the contention nature of Ethernet becomes a weakness is when network traffic is heavy and steady. At peak network usage times within a company, such as 10 A.M., users often experience lengthy delays because of the number of data collisions. Most users, however, simply think that the network is being unresponsive and don't understand the underlying cause.

A second problem with contention networks is that they can't guarantee that a certain user's transmission will take place at a certain time. If a key application needs to be run at 10 A.M. and the results distributed on the network, Ethernet won't likely be the first choice of network topologies because heavy usage will cause delays in all users' transmissions.

The growth and popularity of 10BaseT Ethernet. Ethernet's growth accelerated around 1990 when the IEEE 802.3 Committee formalized a set of specifications for Ethernet over unshielded twisted-pair wire. This set of specifications is widely known simply as *10BaseT*. Its name comes from the particular method used by the committee to describe the bandwidth and distance that data can be transmitted without loss. Ethernet over unshielded twisted-pair wire joined 10Base5 (thick coaxial cabling transmitting for 500 meters) and 10Base2 (thin coaxial cabling transmitting for 200 meters). The general guidelines for 10BaseT are a maximum segment length of 100 meters.

While the ease of installation and inexpensive copper wiring were major reasons for Ethernet's growing popularity, the change in physical structure was also very appealing. Instead of a central cable acting as a data highway, 10BaseT used a wiring concentrator, also known as a *media access unit (MAU)*, or simply a *hub*. Among the advantages offered by a wiring concentrator were more convenient network management as well as greater fault tolerance. "Intelligent hubs" soon became the norm because they provided diagnostic information, the ability to make efficient moves and changes, and a way of keeping a network up and running even when there was a malfunctioning network interface card or break in the cabling.

While the logical structure of 10BaseT is identical to traditional Ethernet's bus structure and the CSMA/CD network access scheme, the physical structure changed dramatically. Figure 1.5 illustrates how this topology has a physical star structure. The wiring concentrator (hub) has cables radiating from it to the PCs connected to it.

Isochronous Ethernet. The IEEE 802.9a standard describes an isoEthernet standard that supports real-time multimedia over LAN. It specifies a signaling layer based on integrated services digital network (ISDN) protocols and uses the existing 10Base-T infrastructure. It doesn't supply more bandwidth, but it does permit the deployment of synchronized bandwidth to avoid the latency and delay common in most other networks. To transmit simultaneous voice, video, and data, network managers need only connect their hubs to their company's private branch exchange (PBX) or Centrex telephone system, as well as to their LAN.

What is significant about isochronous Ethernet is that it's based on current telephony standards. This means that it can interoperate with 155 Mbps (OC-3) and 622

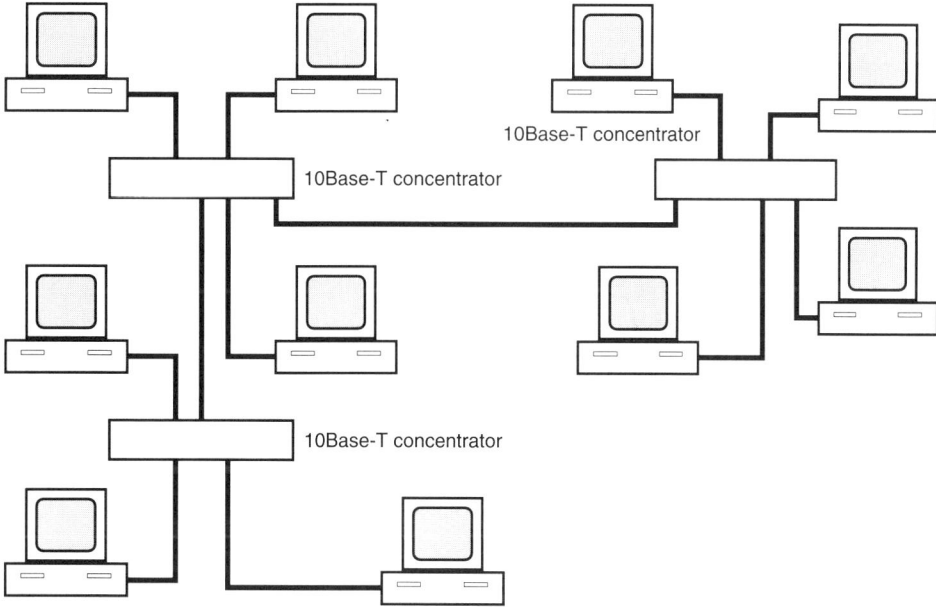

Figure 1.5 A 10Base-T LAN.

Mbps (OC-12), speeds associated with ATM. So isochronous Ethernet might very well play a roll in the future of ATM.

Token Ring

While Ethernet grew and assumed its market leadership position, IBM countered with an alternative topology, Token Ring. While this network architecture was far more expensive than Ethernet, it offered optimized communications with mainframe computers as well as improved network management tools. Perhaps most important of all, it offered an approach that ensured no data would be lost through collisions. IBM's primary customers chose Token Ring for all the reasons just given. It seemed to be a better environment for their crucial applications that had to run successfully at certain times. Figure 1.6 illustrates the Token Ring topology.

Unlike Ethernet, Token Ring is a *noncontention* network. A special bit pattern known as a *token* is circulated across the network. When a network user wants to transmit data over the network, that person's network interface card sets the token's bit pattern to indicate that the network is in use. When the user's message has been transmitted, the bit pattern is reset to indicate the network is not in use. This approach works much the same way as a cab driver who raises the cab's flag to indicate that it's currently in use and unavailable. The flag is lowered when the cab driver is

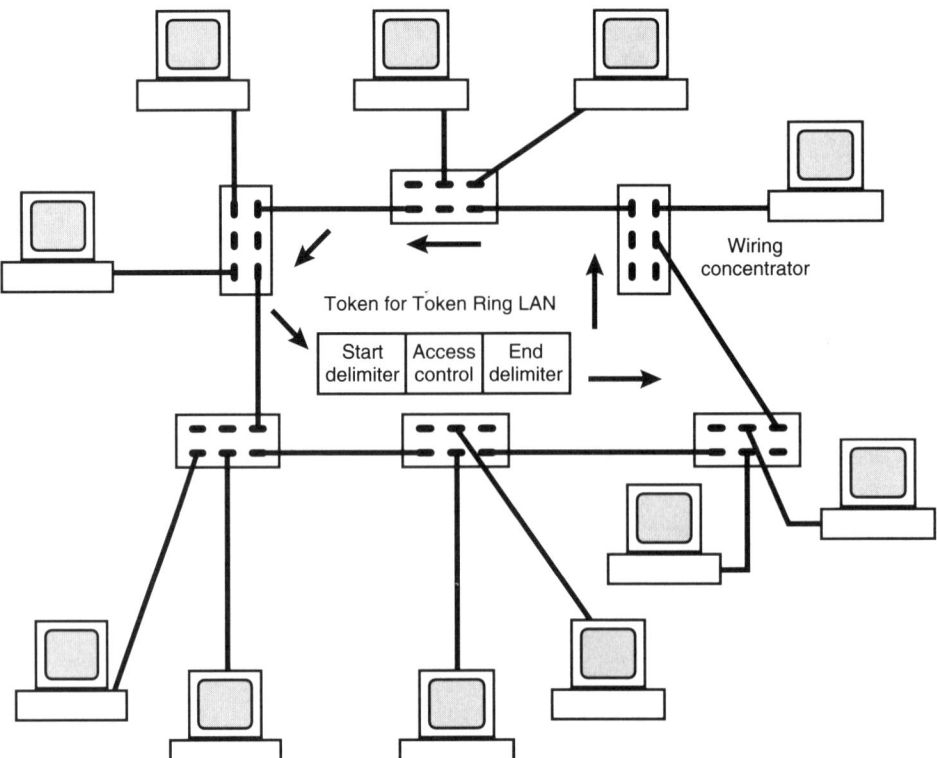

Figure 1.6 A Token Ring network in operation.

TABLE 1.4 Advantages and Disadvantages of Token Ring Topology

Advantages	Disadvantages
Very high throughput	Expensive
Excellent mainframe connectivity	No migration path to higher-bandwidth version
Guaranteed transmission	

ready for another customer. Token Ring LANs transmit data at either 4 or 16 Mbps, depending on the type of network interface cards and wire concentrators. The IEEE 802.5 Committee developed the Token Ring specifications.

Table 1.4 lists the major advantages and disadvantages of using the Token Ring LAN topology. Despite a lower transmission speed than Ethernet, Token Ring networks often enjoy greater throughput compared to Ethernet LANs with very heavy traffic because of its noncontention nature. While Token Ring is far more expensive than Ethernet, it does offer some more sophisticated network management features. It often proves to be a better choice as a network topology for those networks with mission-critical applications that have to be run at certain scheduled intervals. Because IBM specifically developed Token Ring topology for its customer base with mainframe computers, it's a good choice for companies that require extensive LAN-to-mainframe communications. One of the major disadvantages of Token Ring topology has been that it doesn't currently offer users a migration path to a topology with higher transmission speeds. This point will become increasingly important in subsequent chapters when I describe efforts by vendors to entice Token Ring customers with ATM technology.

Fiber Distributed Data Interface (FDDI)

While Token Ring's noncontention approach to network access resulted in higher throughput (a higher overall bandwidth) than Ethernet in some LANs with very heavy data traffic, its maximum bandwidth of 16 Mbps was still not adequate for many corporate networks. The answer appeared to come from the American National Standards Institute (ANSI) in the form of fiber distributed data interface (FDDI), a noncontention topology that provides a bandwidth of 100 Mbps over fiber or twisted pair wire.

FDDI was designed to bear the burden of a large network's traffic as the main data highway. Up to 1,000 stations can be connected to an FDDI network with up to 3 kilometers between stations with the fiber version. It uses a dual ring approach that offers built-in protection against system failure. The primary ring carries information and a secondary ring carries control signals. If a break in one set of cabling takes place, communications can continue through the other set of cabling. The FDDI frame carries a maximum size of 4,500 bytes, making it ideal for large data transfers. Figure 1.7 illustrates a typical FDDI LAN.

While on the surface Token Ring and FDDI seem to be similar, they operate quite differently. As mentioned earlier in this chapter, a workstation that wants to transmit a data packet under Token Ring grabs the bit pattern known as a token, and then transmits a frame down the ring to its destination station. Under 16-Mbps Token

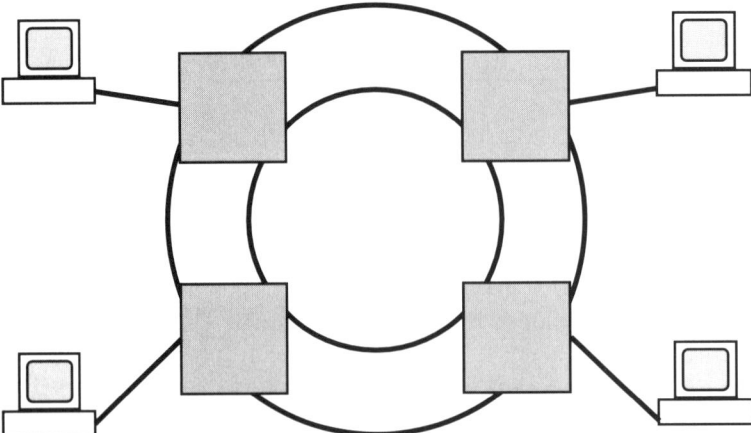

Figure 1.7 An FDDI network.

Ring, multiple frames can be transmitted. When the token and frame returns to the transmitting station, it can transmit a new frame.

Under FDDI, possession of the token is timed according to predetermined limits. While the station has the token, it can transmit multiple frames and then gives up the token for use by another station, without stripping frames before releasing the token. FDDI also offers built-in network management. Each FDDI station has built-in integrated station management (SMT) in the form of either software or firmware built into the network interface card. SMT monitors the network and provides information about the network's current performance and configuration.

Table 1.5 describes the major advantages and disadvantages of an FDDI network topology. It refers to the fiber version, which is far more popular today. While the fiber components are still expensive relative to Ethernet and even Token Ring, the greater bandwidth is appealing. Fiber transmission is immune to electromechanical interference, so FDDI is ideal for environments where this is a problem. The growth of FDDI has been hampered because it contains a different packet structure, which is incompatible with both Ethernet and Token Ring. Many network managers who already have one or both of these topologies installed are hesitant to introduce still another incompatibility problem.

FDDI is also available in a copper version for well under $1,000 per port. Crescendo, acquired by Cisco Systems in 1993, offers a copper implementation known as copper *distributed data interface (CDDI)*. The generic specification is known as *shielded distributed data interface (SDDI)*. Unfortunately for vendors

TABLE 1.5 Advantages and Disadvantages of an FDDI Topology

Advantages	Disadvantages
100-Mbps bandwidth	Expensive
Immune to electromechanical interference	Bridges or routers required for interoperability with more common Token Ring or Ethernet LAN segments

supporting this version of FDDI, the announcement of a 100-Mbps version of Ethernet has seriously eroded interest. Why introduce still another new topology when Ethernet might be the answer? I'll examine this new version of Ethernet later in the chapter.

Other network architectures

When upper management decides that it wants all corporate networks linked together into an enterprise LAN, there's often more than just Ethernet, Token Ring, and FDDI to worry about. LANs using the AppleTalk protocol often link Macintosh computers with Apple laser printers. AppleTalk describes the specifications for how data accesses such LANs as well as the format for its packets. The fields in these packets don't match up with any other topology, so devices known as *routers* (discussed later in this chapter) are required to link such networks with networks using Ethernet, Token Ring, etc. Apple provides a very inexpensive network interface for its products, known as *LocalTalk*. Built into many of its computers and printers, this interface, when attached to shielded twisted-pair cabling, transmits data at 230,400 bps. Apple realizes that this bandwidth is no longer adequate for many applications, however, and now ships most of its new computers with Ethernet interfaces.

The other major network topology found in many corporations is Arcnet. Offering a limited bandwidth of 2.5 Mbps and very limited interoperability with other topologies, but with the advantage of being virtually indestructible, *attached resource computer network (Arcnet)* is one of the oldest LAN topologies, developed by Datapoint in 1977. Currently at the twilight of its sales cycle, it has achieved the status of an ANSI standard. Arcnet is easy to install and easy to maintain. The major problem for network managers is that there are many LANs with this topology that now need to be brought into a corporate-wide network. The Arcnet packet is very different from the topologies described earlier. To many younger network managers it might resemble a creature that somehow survived the evolutionary process. It clearly functions, but what exactly is it?

How Information Travels Over a LAN

Ethernet is the chosen network structure (or topology) in over 60% of all installed LANs, while Novell's NetWare (in various versions) holds between 65% to 70% of the share of market for network operating systems. For the purposes of our example, let's assume that a company has a network that contains these two components. In order to understand how the various LAN building blocks work together, let's assume that Widget Corporation has a local area network that uses an Ethernet network architecture and is governed by the NetWare network operating system. Carol Wilson is a sales manager who wants to send a brief message and some key budget information to Bill Jackson in Accounting. Let's follow the path along which this information will follow and the changes in its structure as it moves from Carol's personal computer to Bill's personal computer.

Carol writes a note to Bill and issues the command to transmit the message. The client portion of the network operating system (NetWare) is running on Carol's machine. It intercepts the message and communicates with the server portion of the NOS residing on one of the LAN's file servers. NetWare uses its own set of software rules, or *protocols*, to format the information. The message is now in the form known as a *NetWare core protocol (NCP)* request packet. As seen in Figure 1.8, NetWare takes this request packet and encapsulates the information into a packet that has a format known as *internetwork packet exchange (IPX)*.

While today's computers and LANs are very powerful and can move massive amounts of information, their electronic nature requires data in a format they can handle. They still think in terms of bits and bytes, rather than sentences or even words. So when I talk about packets of data, I'm still talking about a series of *bits* (binary digits) grouped as *bytes* (eight bits to a byte). If you looked at these bytes, you wouldn't be able to differentiate which bytes were actual data and which bytes were control information. That's why *where* they are placed is so important. Think of a railroad switching engineer who can't read English. If this engineer knows that the third car on the Wabash Cannonball is always the refrigerator car and must be treated differently (switched to another train so it can reach its destination before the contents spoil), then the whole process works smoothly. As long as a LAN packet always packages its information in the same format and same order, then it can be decoded successfully on its arrival.

Figure 1.9 displays a typical IPX packet. This packet now contains Carol's original data as well as such key information as its source and destination addresses. Control information is also part of this packet, and it includes such critical information as the length of the packet, error checking parameters, and transport instructions. Up to 546 bytes of data along with accompanying control information can be included in such a packet.

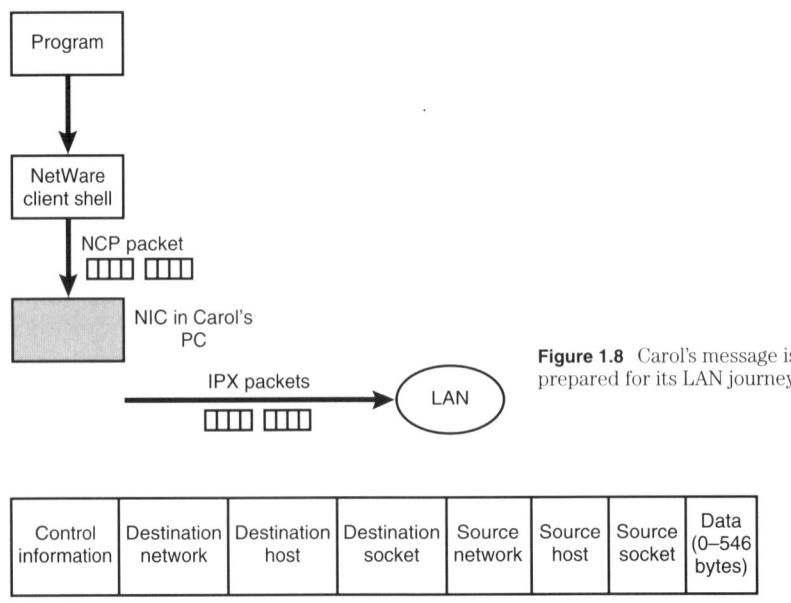

Figure 1.8 Carol's message is prepared for its LAN journey.

Control information	Destination network	Destination host	Destination socket	Source network	Source host	Source socket	Data (0–546 bytes)

Figure 1.9 A typical IPX packet.

Network Operating Systems

As mentioned previously, Carol's LAN is governed by its network operating system, which in our example is Novell's NetWare. Because NetWare is an example of a *Client/Server* NOS, its software runs on the PCs acting as *clients* (users needing help from the server) as well as on the servers. The software running on the client intercepts input/output commands that normally would go to a PC's stand-alone printer, storage, and other input/output resources, and redirects these commands to the file server. The NOS running on the server performs several different functions. It coordinates requests for information to be retrieved from or stored on the server, it manages requests to use LAN printers and other devices, and it provides a variety of management functions such as security and diagnostics. Servers on a NetWare LAN generate considerable traffic with each other as well as with the client software running on LAN users' computers.

Network File Server

Carol's client NetWare software communicates with the NetWare server software stored on a network file server. While minicomputers and even mainframes can function as LAN servers, these servers are most often microcomputers that have been optimized for this task. The optimization of PC models takes the form of a fast microprocessor or even multiple microprocessors, special server diagnostic software, beefed-up storage capabilities and memory (RAM), and a fast bus. Think of a file server's bus as the computer equivalent of a manufacturing company's shipping and receiving dock. The bus determines how fast data can be input into the server and output to peripherals such as printers and external storage, and to PCs on the local area network.

Circuit cards slide into the PC file server's bus to provide additional functionality. A network interface card, for example, usually plugs into a bus slot. One important consideration to remember when I discuss ATM technology is that different topologies require network interface cards specifically designed for a particular server's bus. In other words, if two servers have different buses (EISA and PCI, for example), then the manufacturer of an Ethernet, Token Ring, or even ATM network interface card would have to develop cards for each of these different buses. If your office uses Macintosh computers, an ATM manufacturer would have to develop ATM NICs specifically for that bus or you'd be out of luck. The good news is, as you'll see in subsequent chapters, that ATM products are already available for the Macintosh.

I'm taking the scenic route, with detours to look at the various LAN building blocks, but in reality the trip that Carol's message takes over the LAN is lightning fast. The key concept to understand with network communications is that each key LAN component with enough intelligence to do so usually requires the information in its own format.

Let's look at a non-network message delivery system that also requires information to be repackaged in a variety of different formats to ensure that you understand how common it is to repackage the same information. Imagine for a moment that Carol filled out the front of a rather elaborate corporate internal mailing envelope (Widget

Corp. Internal Communications Only) that contained a department, subdepartment, and mailstop number. She then placed her message, on an IBM-compatible diskette, inside this envelope and placed the envelope inside a larger envelope that contained the conventional street address for Widget Corporation's accounting department. The U.S. Post Office would understand the format of this address and deliver the package to the receptionist at the front desk of that building. She would open the envelope and extract the internal corporate mailing envelope that was now in the format required by Widget Corporation's own internal mail delivery system. The envelope would go into an Out basket, be picked up by the internal mail worker, and delivered to Bill Jackson. Bill would open the envelope and take out the diskette containing Carol's information. Since the information is in the appropriate format for a PC to understand, Bill would simply load the diskette into his computer and read it.

Now let's return to the electronic message that Carol generated. The NOS packaged the message so it can handle it, but it isn't yet in a format that can be handled by the hardware components of the LAN. That's the job of the network interface card.

Network Interface Card

The *network interface card (NIC)* contains much of a LAN's intelligence in the form of computer chips that control the way data is packaged, transmitted over a LAN, received, unpackaged, and decoded. The NIC is also responsible for the smooth flow of data over a LAN as well as error checking and recovery. Imagine that you could reduce yourself in size and become an interested observer walking within the complex NIC environment. The action will be fast and furious, but you're bound to enjoy the LANscape.

Continuing tracking Carol's message for Bill Jackson, the MAC (a sublayer of the NIC, to be discussed in a moment) has taken the IPX packet, incorporated information from the LLC, and repackaged the data and control information for transmission over a specific LAN hardware, in this case an Ethernet network. Let's pause for a few moments and examine the topic of LAN topology and look at some of the alternatives available for network architecture.

Protocol layers dictating a NIC's behavior

Earlier in this chapter I indicated that the Institute of Electrical and Electronics Engineers (IEEE) established an 802 Committee that has been developing LAN standards for several years. The beauty of these standards, which are in the form of protocols (or rules), is that they aren't moving targets but industry-wide agreed-upon specifications for LAN operation; thus, they're firm enough to warrant the initial expense of incorporating them into silicon computer chips. If you take a close look at the chips on a typical NIC, you'll observe three clearly segmented layers of protocols corresponding to the OSI model's data link and physical layers. Figure 1.10 shows that these are the physical, media access control, and logical link control layers.

The error-checking control information included with the packet lets the network interface card receiving the information know that the packet is no good and that it must be retransmitted. Carol's message is now incorporated in an Ethernet packet.

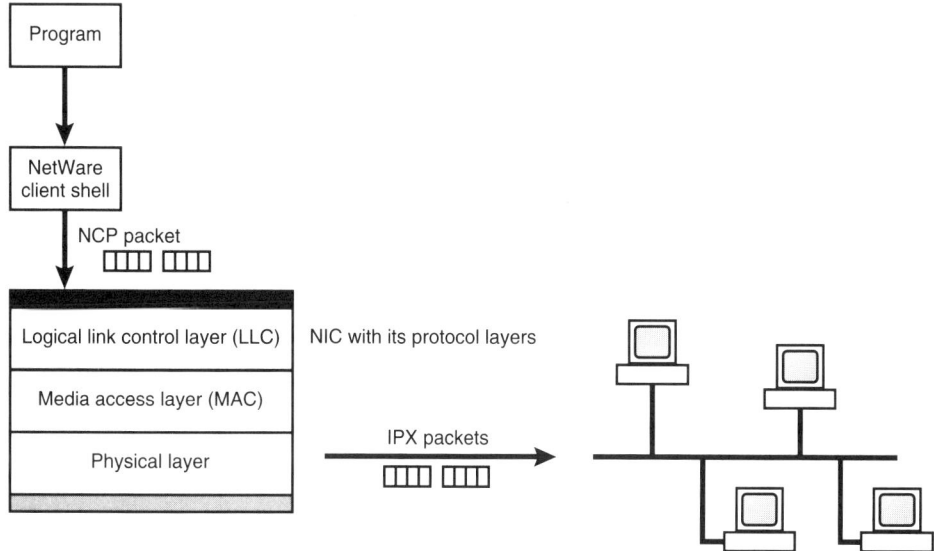

Figure 1.10 An Ethernet NIC's 802 protocol layers spell out how it operates.

The LLC and the MAC

The higher-level protocol layer found within the NIC's chip is subdivided, by function, into two sublayers: the logical link control (LLC) and medium access control (MAC) layers. The LLC layer of protocols within a NIC's controller chip is independent of a specific access method. This means that regardless which topology the NIC supports and how data is transmitted onto the network (contention, noncontention, etc.), the same rules apply at this layer. The LLC, thus, is not hardware-specific. Ethernet, Token Ring, and various other types of LANs all use a common LLC. The responsibilities of the LLC include establishing a LAN connection, transferring data, and terminating the connection. The LAN connection service can be connection-oriented or connectionless. As pointed out earlier in this chapter, these two types of service can be likened to two different classes of customer service: a high quality service with lots of quality control or an express service with greater speed but no real quality control.

Connectionless service is the norm on a LAN because of the inherent high rate of speed and reliability. LANs with this level of service don't use the LLC to provide for error checking, flow control, and error recovery. Instead, they use the appropriate protocols found in the transport layer of the LAN's network operating system. So in the case of Carol's note to Bill Jackson, NetWare will send an acknowledgement when the message is received without error and the NIC won't have to handle this particular task.

The MAC is hardware-specific and responsible for controlling the transmission of a packet of data across the LAN to a destination station or node. It's responsible for providing headers and trailer fields that describe the beginning and ending of a message. It also provides the synchronization necessary to transmit and receive LAN stations in order to communicate and provides error detection. The MAC strips the packets it receives from the layer above it (the LLC) and presents the bit stream to the physical layer folks for electronic transmission. The MAC also receives incoming bit streams from the physical layer folks and packages them into packets the LLC can understand.

Think of the MAC as a factory where raw materials (the bit stream) are received and processed into products a customer (the LLC) can use. Similarly, this NIC factory also takes finished products from the LLC and repackages them so they're capable of undergoing the rigors of being shipped without being destroyed.

Carol's message to Bill Jackson has now undergone several changes. It has been packaged by NetWare and modified by both the LLC and MAC protocol layers found on an Ethernet network interface card. It's now being readied for transmission by the physical protocol layer of the NIC.

The Ethernet NIC's physical layer (or department) has rules to cover how it transmits bits over different types of physical media (thick and thin coaxial cable, fiber, and twisted-pair wire). This layer is concerned with transmitting and receiving the actual electrical signals that correspond to the stream of bits conveying both the data and control information found in an Ethernet packet. If people were working in this department, they would be scientists and engineers who were concerned only with the world of electrical transmission. There's no concern or even interest in the meaning of the message being sent or received.

Carol's message will travel over the LAN as a series of electrical signals. As it passes each user's PC, the corresponding NIC examines the bit pattern that provides the packet's destination address. Imagine a waiter in a posh restaurant who walks around with chalkboard with the name of a person being paged. The diners examine the name on the blackboard and then either ignore it or indicate they're the person being paged. Bill Jackson's NIC will recognize the destination address as its own address and then copy the packet to the memory (RAM) found on the card. Once the packet has been copied, it travels up the various protocol layers of the NIC. In other words, the physical layer translates the electrical signals corresponding to the bit pattern of the packet into a packet that the MAC layer can understand. The MAC layer repackages the information into a format that the LLC layer can understand. The LLC continues the process by repackaging the information into an IPX packet that NetWare can understand. The particular application that Bill uses to read such messages (let's assume it's an e-mail program) would then be able to read the message and report that Bill has an incoming message from Carol.

Of course a message could contain hundreds of such packets, and each packet would follow the path I've just described. The process, however, is transparent to a LAN user. The only time the user would take notice is when there's so much traffic that collisions have slowed down performance. The standard user response is to complain that the network "is slow today." The network is no slower than it was before, but the collisions have created a situation in which NICs are forced to retransmit their packets several times.

Bridges

When a LAN is small and contains only a few users, then traffic is usually very manageable. The messages generated over an Ethernet LAN are broadcast over that entire network. When traffic grows heavy, though, the LAN resembles Los Angeles freeways at rush hour. The traffic congestion from so many NICs attempting to transmit their data slows throughput to unacceptable levels. One solution to this intolerable situation is to separate or segment the LAN into several subnetworks so traffic destined for a specific segment remains on that segment and isn't rebroadcast over all other segments. Only traffic specifically addressed to a node on a different segment can cross from one segment to a second LAN segment. The tool for accomplishing this task is known as a *bridge*.

A *bridge* consists of the hardware and software required to link together two different LANs or subnetworks located at the same site into one internetwork. The simplest type of bridge examines a packet's destination address field and compares the address with a table that lists the addresses of workstations on its network. If the address doesn't match any address on this table, it knows that the packet is not designated for a node on that network. It then forwards the packet to the segment it bridges. This type of bridge, one that checks addresses and then forwards packets meant for a different segment, is known as *transparent bridging*.

Think of this kind of bridge as a not-too-smart receptionist responsible for a particular department. When someone asks to see Ms. Kathy Burns, for example, the receptionist checks a roster to see if Ms. Burns is a member of that department. If her name is not there, the receptionist points to a corridor that leads to the next department. "Maybe you'll have better luck there," says the receptionist.

A more sophisticated bridge creates its own network address tables. It examines the source and destination addresses of every packet transmitted to the LANs to which it's connected. It then builds its own address tables using the subnetwork number and node source addresses of packets it receives. The bridge then tries to match a destination address when it receives a packet. When it matches an address, the bridge *filters* the packet, sending it on its way across the current subnetwork to a destination node that will recognize its own address and copy the packet to its RAM. If there's no match, then it *forwards* the packet across the bridge to the next subnetwork.

Bridges aren't terribly smart. They don't concern themselves with higher-level protocols. They function at the media access sublayer of the OSI model's data link layer, far removed from upper-level protocols. As long as the networks on both sides of the bridge adhere to IEEE 802.2 logical link control (LLC) standards, the bridge can span them regardless of differences in their media or network access method.

While I've been focusing on Ethernet LANs, Token Ring LANs can also benefit from being bridged. One Token Ring subnetwork could use inexpensive unshielded twisted-pair wire, while another subnetwork in an area where there's electronic interference could use more expensive fiber-optic cabling. These two Token Ring LANs can then be bridged into a single corporate LAN. Generally, Token Ring networks are bridged by two-port bridges.

Bridges offer a variety of advantages for the network manager. You've seen that congestion can be minimized and throughput increased, and different media can be used where appropriate. Bridges also add security. They can be programmed to forward only those packets that contain certain source or destination addresses, so only certain nodes can send information to or receive information from another subnetwork. The accounting subnetwork, for example, can have a bridge that permits only certain nodes outside the network to receive information. Finally, bridges can increase the distance a network can span. Since a bridge rebroadcasts a packet to the nodes on the receiving side, it functions like a repeater to increase the distance a packet can travel without its signal attenuating (weakening).

Routers

A *router* is a device that operates at the network layer of the OSI model to find the optimum path a packet should take to reach its destination. Local or remote, routers are ideally suited for large networks where there might be several different paths from one LAN to another LAN. Unlike bridges that don't concern themselves with higher-level protocols, routers are protocol-specific; they're designed to support specific protocols, such as NetWare's IPX, Unix's TCP/IP, or Digital Equipment Corporation's DECnet, and use the appropriate addressing schemes, error checking, and routing techniques that characterize these protocols.

Figure 1.11 shows how routers operate. Unlike most bridges, routers can maintain several alternative paths and select the most appropriate path given certain defined conditions, such as traffic congestion. In this case, a packet indicates a workstation on network XYZ as its destination address. Router 1 looks over its routing tables and determines that the optimum path is through router 4. Unfortunately, router 1 also realizes that the direct path to router 4 is congested. It chooses the alternative path through routers 2 and 3.

Routers are far more intelligent than bridges, and they use this intelligence to determine the optimum path for connecting together two LANs. A router is also smart enough to perform the packet segmentation and reassembly required to accommodate intermediate networks in which packet sizes are different.

One major advantage of a router over a bridge is that it doesn't automatically replicate all broadcast messages. This means that if a device begins to flood a network with copies of a single packet, the routers are able to keep the problem local by presenting a "firewall" that prevents the storm of packets from engulfing the entire network.

Routers use a number of algorithms to do their jobs. Berkeley-derived UNIX systems use routing information protocol (RIP) to calculate how many hops through other routers different paths would encompass. They then choose the path with the

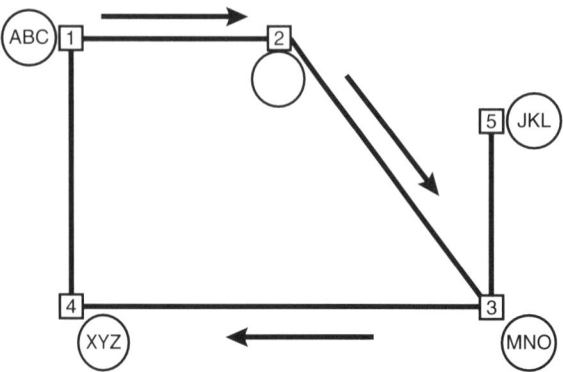

Figure 1.11 A router selecting the best path.

fewest hops as the optimal path. Most routers broadcast packets describing their own links to other routers. All routers on the network use these broadcast packets to assemble their own routing tables.

The U.S. Department of Defense's Internet Activities Board sets internet policy for TCP/IP users. The board's Internet Engineering Task Force (IETF) created the open-shortest-path-first (OSPF) working group to develop a dynamic routing protocol for TCP/IP that would provide features not offered by the routing internet protocol.

RIP is limited when used with networks of more than 100 routers because it frequently broadcasts its entire routing table. On large internets with over 100 routers, routing updates take longer and longer and consume increasing amounts of bandwidth. Also, packets can't travel through more than 15 routers from sender to receiver. This protocol selects a single path to each destination and isn't capable of considering factors such as traffic congestion, delay, and bandwidth.

The OSPF protocol uses a shortest-path-first algorithm. Each router broadcasts a packet that describes its own local links. Routers collect information from these broadcast packets to build their own network routing tables. Since the packets describing local links are very short, they cause far less traffic congestion than RIP's approach of broadcasting very large routing tables describing the entire network.

Another advantage of OSPF is that network managers can configure their routers to provide least-cost routing according to whatever criteria these managers define as a cost. Unlike RIP, OSPF doesn't limit the number of routers that can be used, nor does it limit routing to a single path; loads can be distributed over several different paths to optimize available bandwidth.

Source-Routing Bridges

In 1985 IBM introduced source routing with its Token Ring topology. Source routing is a way for a node to solicit information that helps it create a roadmap to a destination node so it can attach this roadmap to the data it wants to transmit. Source-routing bridges perform routing duties at the network layer of the OSI model. Each device on a LAN that uses source routing must have a unique six-byte address. The address uses one bit (the I/G bit) to indicate whether there's an individual or group address.

Source routing takes the I/G bit in the source address only and uses it as a routing information (RI) indicator bit. When this bit is set to 1, it indicates the presence of additional routing information in the frame header. This additional routing information specifies the frame's complete path from source workstation to destination workstation.

Each LAN ring is assigned a unique number just as an office routing slip might indicate the order in which a memo should be circulated before it's filed.

A networked PC on a Token Ring network gathers routing information by transmitting an all-routes broadcast frame to all rings connected on an internetwork. This frame contains control information as well as a blank buffer to be filled in by other nodes. Bridges fill in the numbers for the two rings they connect and their own bridge numbers. The destination node receives this broadcast frame and returns it to the source station, which now has road map of the route that the frame took.

Bridges or Routers?

Bridges are ideal when two networks with different higher-level protocols but the same MAC layers need to be linked together. Bridges are relatively inexpensive and much faster than most routers. They're also much easier to install and maintain. Once installed, bridges can automatically learn the network location of stations by listening to the source addresses of network traffic.

Bridges are not ideal, though, for large, complex networks—for a variety of reasons. Since bridges pass all traffic, including broadcast storms (a malfunctioning NIC that broadcasts constantly), a few NIC problems could bring down a very large network. Also, since many bridges require a single path between networks, they lack the system fault tolerance (recoverability from system failure) that routers' multiple paths provide.

More and more networks are now running multiple protocols, so a major advantage of a router is its ability to pass packets with specific protocols from one network to another. Routers can adjust to changing network conditions and provide network management functions not offered by bridges.

Another major advantage of a router over a bridge is its ability to perform packet segmentation and reassembly in order to accommodate intermediate networks in which packet sizes are different. An example of this situation is a connection between two Ethernet networks running NetWare via an Arcnet network running NetWare. Ethernet and Arcnet packet sizes vary considerably.

While they're much more complex and expensive than bridges or routers, some situation might require a hybrid of the two devices, called a *brouter*. A brouter is a hybrid router and bridge that can perform the functions of each. It first attempts to make a routing decision, but reverts to bridge status if unable to do so.

How LANS Are Used

LANs are used a number of different ways in the business world. Because ATM products are currently being designed for several different types of LANs, let's examine the various ways these networks can be used.

Backbone LANS

A human backbone goes right down the center of a person's back, providing protection for the vital communication taking place along the spinal chord. If the backbone and hence the spinal cord are injured, the communication fails. In much the same way, a LAN backbone serves as the central communication highway to link together several different network segments.

Often a backbone contains no nodes, just bridges linking it to several different networks. One way to think of a backbone is as a network of networks or as a gigantic high-speed communications switching center. Figure 1.12 describes a LAN backbone linking together several Token Ring LAN segments.

Departmental LANs

A departmental LAN generally revolves around key departmental applications that all users share. Accounting is a good example of a department that requires virtually no access by non-Accounting personnel. Isolating the Accounting department's ap-

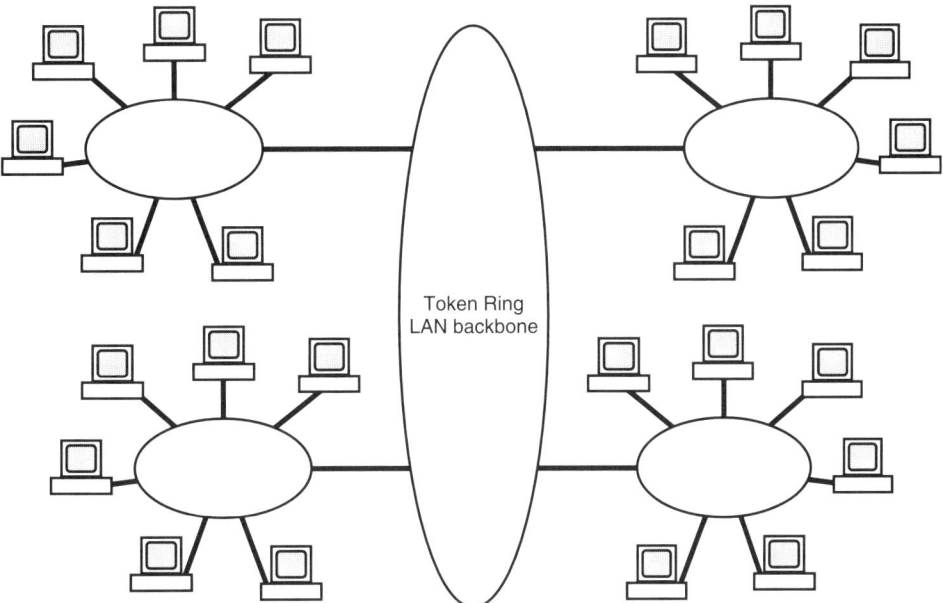

Figure 1.12   A LAN backbone links together several Token Ring LAN segments.

plication on a LAN segment of its own provides greater security since the packets are not broadcast to all corporate LAN users. What appears to the naked eye to be one large corporate network in reality might consist of five or more departmental LANs, all linked together via bridges or routers.

Workgroup LANs

Sometimes the LAN traffic emanating from a particular workgroup is so heavy that it's smart to take that group and put it on its own LAN segment. Engineers who work with CAD/CAM/CAE (computer-aided design, manufacturing, or engineering) applications often generate very detailed graphic designs as well as mountains of data. It makes sense to give these groups their own LAN segment so they don't congest the entire LAN. Of course, even with their own segment, several users generating huge graphics files could congest that segment. One solution, as you'll see shortly, could be an Ethernet switch.

Enterprise networks

The strategic direction most companies are taking today is toward linking all corporate networks together into an enterprise network. Such a network often incorporates different LAN topologies, high-performance workstations, minicomputers, mainframe computers, and remote offices.

Different Technologies to Increase Bandwidth

As mission-critical applications (programs upon which the survival of a company depends) continue to be moved from mainframe computer platforms to LANs, the need for greater bandwidth increases. The most common mission-critical application migrating to LANs today is the company customer database. These databases often generate very large customer records that must be transmitted over the network, thus increasing the demand for LAN bandwidth. Similarly, new LAN applications are beginning to make more use of graphics, bit-mapped images, and even video.

When Ethernet's bandwidth usage rates approach 35 percent on a continuous basis, its CSMA/CD approach to network access begins to break down as the network slows. If the problem appears to be limited to a few high peaks of traffic demand for relatively short periods of time, the solution might be as simple as upgrading certain key servers with one of the technologies described in this section. If usage stays high for long intervals (a few minutes), the problem might be that the backbone needs upgrading.

Network managers are beginning to look for new technologies that can increase their current bandwidth. There are many candidates for the task. Let's examine several of these competing products.

Full-duplex Ethernet

In 1993, Kalpana introduced a full-duplex Ethernet technology. It consisted of two 10-Mbps channels, one for receiving data and one for sending data in a point-to-point connection. Both ends of a full-duplex connection can simultaneously send and receive data via a null model cable, resulting in a maximum aggregate of 20 Mbps. This technology is now available from a number of vendors.

There's a major performance limitation, however, to full-duplex Ethernet. The only way this technology can achieve close to a 20-Mbps speed is when traffic is balanced in both directions. Since most client/server communication is primarily one way, it's likely that overall performance will fall below expectations. Cards using this technology do provide far higher throughput even at half-duplex mode, however, so network managers should still consider full-duplex Ethernet as a tool for improving overall network efficiency.

While several vendors offer this technology (IBM, Kalpana, Cabletron, etc.), there's a lack of interoperability among their various products. Unless a customer is already locked into a specific vendor's proprietary line of products, it's probably wise to think twice before committing resources to this technology because it doesn't offer the interoperability that most network managers seek for enterprise-wide networks.

Ethernet switching on a LAN

The years 1990 to 1995 saw a sensational growth in network hubs. The traditional Ethernet backbone began collapsing into the hub's backplane, each hub circuit card or module was able to support an entire LAN segment, and several of these LAN segments were linked via a bridge or router module within the hub. Despite the elegance of getting rid of the backbone cabling, this hub still had the traditional limitations associated with Ethernet. Each node on a LAN segment is still limited by CSMA/CD to an aggregate bandwidth of 10 Mbps. If several users attempt to access the LAN, then collisions reduce the theoretical 10-Mbps throughput of the network to just a few megabits per second. Every node on a LAN segment is subject to the broadcast of packets for all other nodes on that segment. In effect, each node has to look through every other node's mailbox in order to find its own mail. Similarly, if one workstation broadcasts a very heavy data stream, it can choke the entire LAN segment.

Let's see how the situation improves if you make the hub a "switching hub." Now the LAN switch is a very intelligent traffic cop, one who makes networks run with much greater efficiency and with optimized bandwidth. Data passes through a silicon-based switching fabric to create the equivalent of a telephone system in which temporary dedicated "virtual" circuits are created, linking various switch ports. These circuits remain in existence only long enough for a packet to be transmitted to its appropriate destination. Each packet within the Ethernet switching fabric has a destination address. This address determines which port the packet will exit. Enhanced security is the result since no other ports except the ones that the packet enters and leaves has a trace of its existence.

A switch reads the source and destination addresses on a packet but doesn't have a router's ability to convert protocols or select alternative paths for packets to take. Switches do have a major advantage over bridges because they can switch packets among multiple Ethernet segments rather than merely between two segments.

A LAN switch creates physical and/or logical LAN connections that permit several devices to communicate at the maximum LAN speed concurrently. Instead of only one user being able to access an Ethernet LAN, several users can transmit simultaneously. Figure 1.13A shows a traditional Ethernet backbone with three LAN segments,

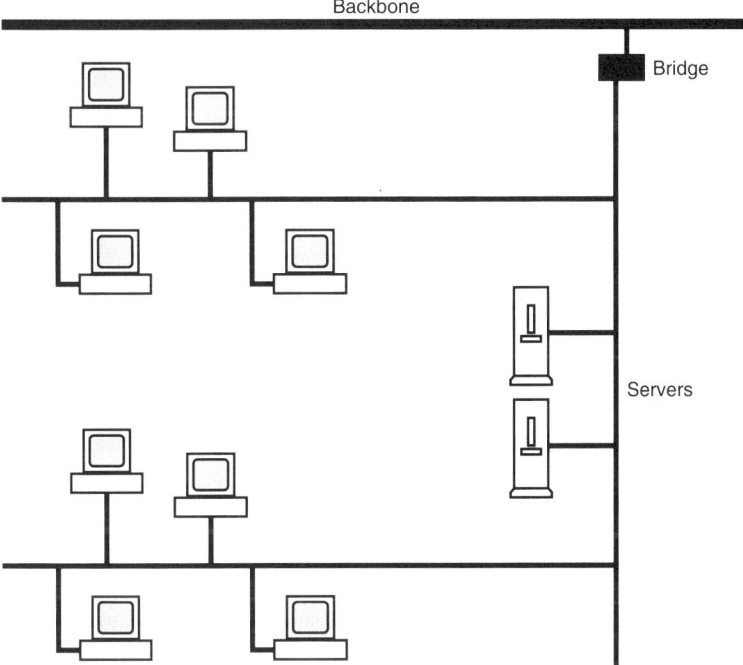

Figure 1.13A A traditional Ethernet backbone and LAN segments.

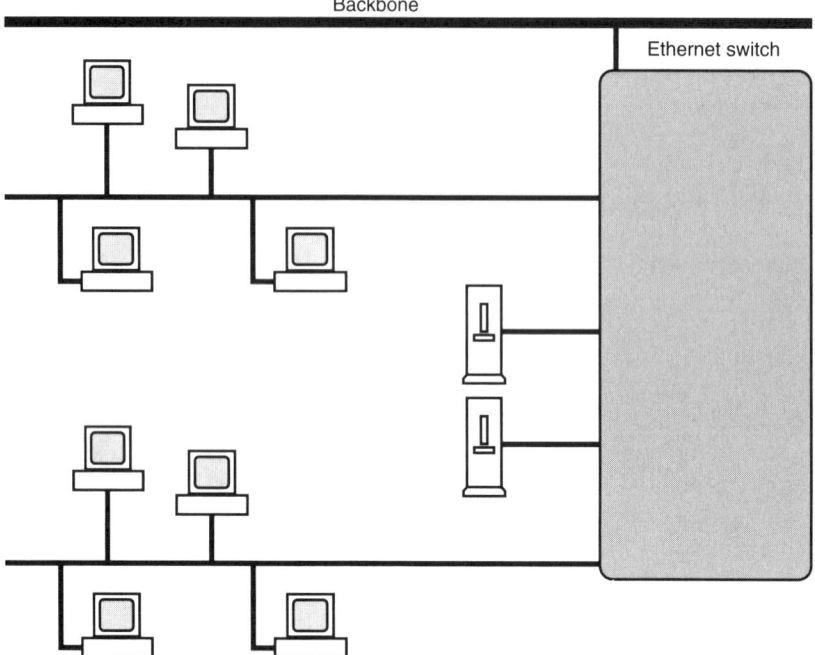

Figure 1.13B   An Ethernet switch replaces the original design.

two servers, and a bridge. Only one user at a time can access the 10-Mbps backbone bandwidth. In Figure 1.13B, a LAN switch has been installed as a kind of collapsed backbone. Now each LAN segment as well as each server have access to 10-Mbps bandwidth concurrently for an aggregate bandwidth of 50 Mbps.

Ports are assigned to intelligent backplanes under software control. Shared memory takes the place of a shared bus given a fast switch performance. The switch can use a cut-through or store-and-forward switching. *Cut-through switching* enables the switch to forward packets without having to wait for the reception of a complete packet. It accomplishes this task by reading the destination address header field on a packet and then forwarding the packet to that address. The efficiency of this switch depends on the type of protocol and size of the packet transmitted. This particular approach is very effective when small packets are being transmitted. NetWare IPX packets, for example, can be switched optimally with this type of switch.

The *store-and-forward* approach to switching uses address tables to keep track of devices linked to a switch's ports. The switch reads the entire packet and then forwards it only if the packet has been received completely and it is error-free. There's a certain amount of overhead in the form of buffered memory associated with this method, but a major advantage is that packet filtering and routing can take place. A network manager can specify that certain network nodes (by address) be treated as the same network segment. Similarly, routing can take place.

Still another advantage of the store-and-forward approach is the ability to switch traffic between isochronous segments. This means, for example, that incoming LAN segments can be based on 10-Mbps Ethernet technology while the outgoing LAN segment can run at full duplex Ethernet (20 Mbps) or even at fast Ethernet (100 Mbps).

The best of both worlds

A new type of switch seems to be emerging that combines the best features of cut-through mode (speed) and store-and-forward (error checking). Cisco's Kalpana division uses what it calls an *error-free cut-through* mode for its Prostack switches. These switches use specialized silicon on each port to enable them to read the address header as well as the CRC (cyclic redundancy check) error-checking information to ensure data integrity. Each port uses its ASIC (application-specific integrated circuit) to store just enough of a packet to read its address and then immediately forwards it to the appropriate outgoing port. As the end of the packet passes through the incoming port, the ASIC acts in cut-through mode to read the CRC. If it finds a bad packet, it reverts to store-and-forward mode and buffers incoming packets, inspecting them for errors and transmitting only good packets. IBM uses much the same technique in its Token Ring switches.

One way of differentiating switches is by their *latency*. The latency of a switch is measured from the time the first bit of a packet is received on one port to the time the same bit is transmitted on a second port. With cut-through mode, a switch's latency is independent of packet size because a forwarding decision is made as soon as the source and destination addresses are read from the packet. In store-and-forward mode, latency is equal to the packet reception time plus the forwarding decision time. So the total latency is the time, depending on packet size, plus a certain amount of fixed overhead time.

In the case of Ethernet switches, there's an additional way to differentiate these products, which is how they handle flow control. If a switch's port contains a number of nodes and these nodes begin transmitting data in excess of the aggregate capacity of that port (10 Mbps), then data would be lost as packets are dropped. To avoid this situation, some switches offer what's called *back pressure*. This technique detects that data is overloading a specific port and generates a collision-detection signal that fools the port into thinking there has been a collision. It therefore backs off and forces nodes and segments to wait before retransmitting data.

Converting existing Ethernet hubs into switching hubs

Switching hubs are often installed where traditional Ethernet hubs currently exist. Some vendors such as 3Com and Madge permit their conventional Ethernet hubs to be upgraded to switching hubs by upgrading the hub-repeater cards with hub-switching cards. Other vendors offer switching hubs that contain external ports that can be connected to an Ethernet segment on an already installed conventional hub. While using a switch's ports to link segments is the most common application today, there are environments where the traffic generated by specific workstations or servers is so heavy that they require their own switching port.

Switched Token Ring

While Token Ring is a noncontention network architecture, many companies are finding they need more bandwidth throughput than 16 Mbps for some of the traffic-intensive applications they plan to add to their LANs. An equally compelling reason is the added efficiency and economics of being able to eliminate two-port bridges that are currently used to connect different ring segments into a corporate-wide LAN. Figure 1.14A shows traditional Token Ring LAN architecture with two-port bridges, compared in Figure 1.14B to switch-based Token Ring architecture. Multi-port routers, the only other option to replace two-port bridges, can be prohibitively expensive.

Just as in the case of Ethernet switches, Token Ring switches come in cut-through and store-and-forward varieties. The traditional sporadic, burst-like traffic associated with LAN workgroups can best be handled by a cut-through switch, while store-and-forward Token Ring switches are better candidates for very heavy backbone traffic. Some Token Ring switches offer transparent bridging by learning the addresses of Token Ring nodes and mapping their addresses to a specific port on the switch.

Some Token Ring switches already take advantage of *dedicated Token Ring (DTR) connections* as a way of increasing bandwidth to nodes generating heavy traffic. This new access method, to be included in the IEEE 802.5 standard, is used in full-duplex mode and removes the token and ring repeat path so that nodes can transmit and receive data simultaneously. Unfortunately, full-duplex mode requires new network interface cards able to perform this task. Half-duplex DTR uses the

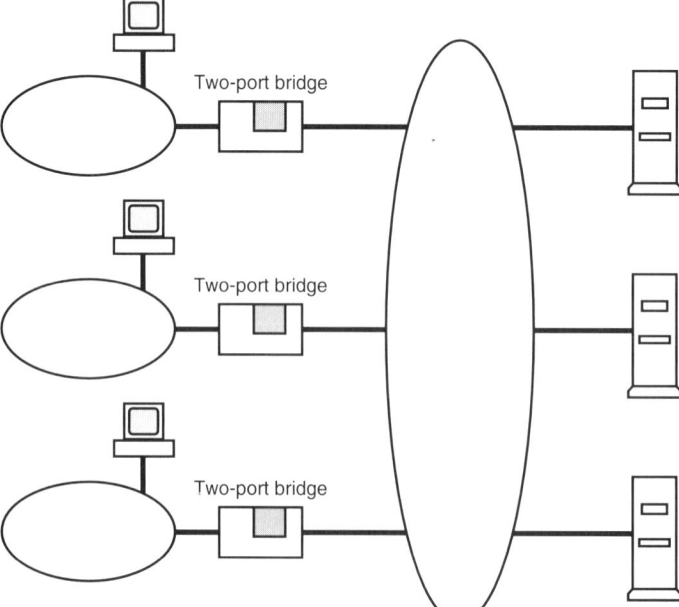

Figure 1.14A Traditional Token Ring architecture.

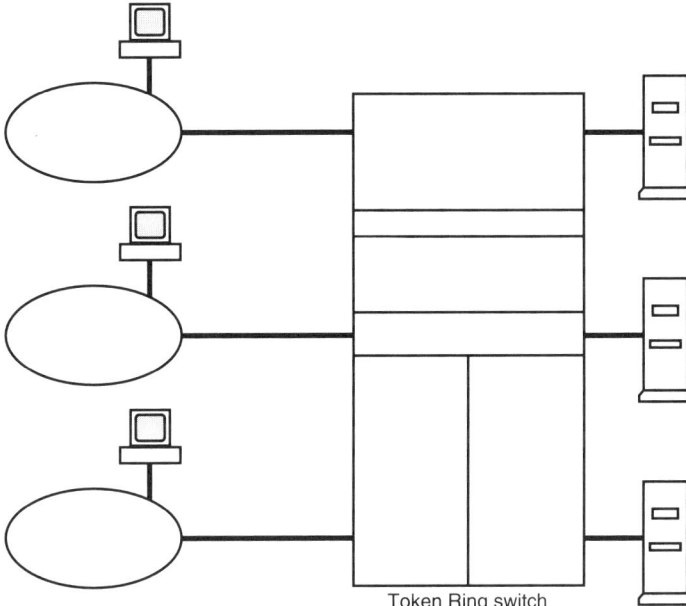

Figure 1.14B Switch-based Token Ring architecture.

conventional Token Ring access method and doesn't require new NICs. The token is transmitted back and forth between the switch and nodes that want to transmit data. Half-duplex DTR permits nonstop 4-Mbps or 16-Mbps transmission, depending on the NICs that are installed.

Token Ring switching is particularly attractive in large networks because of the design problems network managers now face in these environments. IBM's source-route bridging (SRB) doesn't have the flexibility to permit rerouting packets around congested links. Even more crucial, SRB permits only a maximum of seven bridges or "hops" across a network between two communicating end stations. To counter this problem in large networks, managers have installed a hierarchical structure with departmental LANs and servers flowing into a single token ring that serves as a campus backbone. The problem is that increased traffic at the departmental level has forced managers to increase segmentation at that level and thus increase the hop count. As the network continues to grow, the lack of scalability in Token Ring networks creates more and more problems as the backbone begins to carry an increasing load of traffic.

The beauty of Token Ring switches is that where dual port bridges were used, managers can install switches and have these switches replace the physical backbone ring as well as all of the bridges. Any two ring segments can then benefit from full 16-Mbps links rather than having to share this bandwidth with all other rings on the backbone.

As you'll see in the next chapter, there's a close relationship between Token Ring switches and ATM. Token Ring switches need a high-speed interconnect so large volumes of traffic can be carried between switches. While FDDI is a candidate for this role with its 100-Mbps bandwidth, ATM's 155-Mbps bandwidth is even more appealing because of the features associated with ATM. In the next chapter, you'll learn how ATM is capable of LAN emulation.

100BaseT

When Grand Junction first announced that it was developing a 100-Mbps topology known as *fast Ethernet*, it seemed too good to be true. It would use the same CSMA/CD technology and packets, the company promised. This meant 100% compatibility with existing software. The reason 100BaseT is able to provide a bandwidth of 100 Mbps is that it boosts the speed of transmission by a factor of 10 by increasing the bit rate, making the bit times shorter, as well as reducing packet overhead and cable delays.

In this book I'll keep the description of fast Ethernet generic by referring to 100BaseT. The 100BaseT4 specification refers to four-pair wire, category 3, 4, or 5 unshielded twisted-pair wire (UTP). The 100BaseTX specification refers to two-pair, category 5, unshielded twisted-pair wire, as well as category 1, shielded twisted-pair wire (STP). There's also a specification for 100 Base-FX for two strands of 62.5/125 micron fiber.

One key feature of 100BaseT is *auto-negotiation*. Because fast Ethernet products are ultimately likely to support 10 Mbps, 10-Mbps full duplex, 100 Mbps, and 100-Mbps full duplex, an auto-negotiation algorithm permits network adapters and hubs to differentiate the speed of transmission as well as between half- and full-duplex transmission. It also allows automatic detection of cabling types and configuration of a connection to the highest speed supported by both devices.

100VG-AnyLAN

This is a 100-Mbps technology originally developed by Hewlett-Packard and actively promoted by IBM. Because it's a noncontention version of Ethernet, the IEEE turned over development of specifications for this standard to its 802.12 Committee rather than its 802.3 Committee. The 100VG-AnyLAN technology requires a new medium-access protocol layer called *demand priority access method (DPAM)*. A major advantage of DPAM is that it permits some prioritization of time-sensitive traffic such as real-time voice and video. Vendors are positioning this version of fast Ethernet as a way of handling the network traffic demands imposed by multimedia applications. A second major advantage of DPAM is that it supports both Ethernet and Token Ring.

This version of fast Ethernet requires data-quality cabling. It requires four-pair voice-grade (VG), category 3, 4, or 5, unshielded twisted-pair wire. Up to 100 meters of cabling distance is supported for four-pair, category 3, 4, or 5, unshielded twisted-pair wire.

The 100VG LAN physical topology consists of scalable stars. Figure 1.15 illustrates a typical network. While both Token Ring and Ethernet can be supported, they can't exist on the same LAN since they use different frame formats. A router must be used.

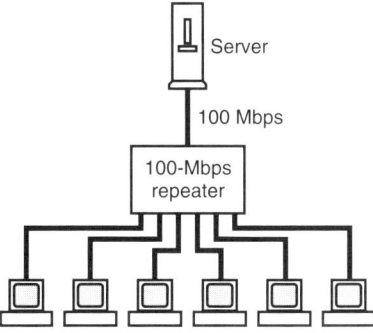

Figure 1.15 A typical 100VG LAN.

Under the demand priority access method, a node that wants to transmit data over the network places a request with the LAN hub or switch that services the nodes making such requests sequentially. DPAM functions in the most democratic manner. All nodes that make requests to transmit are polled in a round-robin fashion and then allowed to transmit a single packet. The process then repeats itself. This process gives all nodes an equal opportunity to access the LAN.

DPAM permits certain applications to be labeled as high priority. High-priority traffic is then sent before the rest of the traffic. Specific ports such as those to servers can also be designated to operate in high-priority mode. If several requests are all designated as high priority, then DPAM services them in turn before servicing the normal-priority requests. If there are so many high-priority requests that normal requests are not being serviced in a timely manner, DPAM will raise their priority to increase the likelihood that these requests are handled before they time-out.

100-Mbps switching hubs

An increasing number of vendors are incorporating both 10-Mbps and 100-Mbps ports in their switches. This gives network managers much more control in providing greater bandwidth precisely where it's needed. In many cases, they're allocating the 100-Mbps ports to their busiest file servers for server-to-server traffic.

Which is better: 100VG or 100BaseT?

This is a good question, but it's not an easy one to answer. Let's examine the advantages and disadvantages of each topology. 100VG can support both Ethernet and Token Ring, which is clearly an advantage for companies that have both topologies installed and are looking for a migration path. Over half of Fortune 1000 companies find themselves in this position according to Computer Intelligence InfoCorp. A second major advantage of 100VG is that since there are no data collisions, throughput is bound to be much higher than with 100BaseT. Finally, 100VG offers rudimentary traffic prioritization. This feature is important to companies that want to run multimedia and real-time video applications over their LANs.

There are, however, several negatives associated with 100VG. The major disadvantage is that it isn't Ethernet. The packet structure isn't identical to Ethernet and thus introduces an additional element of interoperability into an enterprise environment. While every major LAN operating system and application has hooks to conventional Ethernet and CSMA/CD, no drivers and protocols have to be developed for 100VG. Finally, the vast majority of vendors have opted for 100BaseT. This means that prices are bound to remain higher for 100VG while support will probably be less available.

The 100BaseT topology has the major advantage of offering 10/100 scalability. This means that users can buy switchable cards, continue to use conventional Ethernet, and then upgrade at their convenience. Unlike 100VG, all current Ethernet applications will run without change on 100BaseT. Finally, it's possible that the data-quality cabling requirement for 100BaseT is an advantage rather than a disadvantage.

Hewlett-Packard argued that this topology was more economical because it could make use of existing voice cabling. The assumption, of course, is that network managers will want to use 100VG for their entire backbone. It's more reasonable to assume that 100-Mbps Ethernet will be implemented on a segment-by-segment basis wherever it's needed. If this is true, then the savings on not requiring data-grade cabling is minimal.

In short, 100BaseT appears to be the more viable 100-Mbps technology. Having said that, I must also point out that this technology will probably be best used at the workgroup level. Using a 10/100 intelligent hub allows a network manager to take the major burden found on many LANs, the traffic generated among servers, and remove it by providing 100-Mbps links. The rest of the traffic can still run on conventional Ethernet.

A company making a strategic planning decision about their network would have to be very sure that its network traffic growth would be minimal before committing to 100-Mbps Ethernet as a backbone topology. The prices for ATM switches are dropping rapidly, so the prospects of having a backbone that can start at 155 Mbps and be scaled up to 620 Mbps or even 1 Gbps is very appealing.

Fibre Channel

With support from the American National Standards Institute (ANSI) as well as both IBM and Hewlett-Packard, *Fibre Channel* is an emerging standard for high-speed transmission of data over fiber-optic cable at speeds exceeding one billion bits per second. Its major advantage beside speed is that it's a nonblocking switching technology, which means that multiple communications can take place without any collisions.

Fibre Channel was designed originally as a high-speed network architecture for connecting both traditional network devices such as PCs and workstations and high-speed hardware that's traditionally connected directly to system buses, such as hard disk drives, buses, and channels. This technology currently supports four transmission speeds: 133 Mbps, 266 Mbps, 530 Mbps, and 1.06 Gbps. ANSI has approved the standard for 2.134-Gbps and 4.25-Gbps Fibre Channel networks.

A Fibre Channel network performs switching by having ports log directly onto each other or to connecting devices known collectively as *the fabric*. Three classes of service are currently provided:

Class 1. Provides a high-throughput connection for extended periods of time, as might be required by real-time imaging applications.

Class 2. Provides shared bandwidth. Multiple frames can be multiplexed across a connection composed of one or several channels. This class doesn't guarantee delivery of frames nor the order in which they're delivered. It's acceptable if a transport protocol such as Novell's SPX protocol is used that does guarantee delivery information.

Class 3. Is similar to class 2 but confirms frame delivery. As a result, this class of service requires buffers at both ends of the connection. This class is suitable for time-sensitive applications where a dropped frame is useless if it isn't received in time.

It's no secret why IBM and Hewlett-Packard are behind Fibre Channel. Both companies have invested research funds in developing products using this technology for both the minicomputer and mainframe worlds. Hewlett-Packard currently offers high-end workstations that support Fibre Channel. There are network interface adapter cards available for EISA and PCI-based PC systems. Compaq plans to use Fibre Channel to connect clustered servers and disk farms sometime in 1996.

Network managers, however, must consider some of the negatives of Fibre Channel before implementing it. Because this type of network is new to the world of PC LANs, there are no routers available to link existing Ethernet and Token Ring LANs. Equally important is the question of whether Fibre Channel is a dead end that will be made obsolete as ATM products proliferate.

A second obstacle to Fibre Channel is that in a fiber-based, multilink, "switched" topology, high-speed transmission channels are set up very much like telephone calls, which are switched through a telephone exchange. What this means is that Fibre Channel could take as long as 10 seconds to actually set up a high-speed connection. This time delay means that Fibre Channel could be used to link optical devices to network servers, but probably wouldn't be useful for network links, which require real-time access.

Summary

The open systems interconnect (OSI) model is a set of software specifications that describes how layered protocols work to enable two networks to communicate. A local area network is a communication network used by a single organization over a limited distance that permits users to share information and resources. The basic building blocks of a local area network are the network operating system (NOS), the network interface card (NIC), media (cabling), and file servers. Network architecture (topology) can be defined as contention (competing) and noncontention (noncompetitive) access to a LAN. Ethernet offers 10-Mbps bandwidth, but heavy, continuous data traffic can reduce throughput to less than 2 Mbps. The Token Ring topology provides either a 4-Mbps or 16-Mbps bandwidth but uses a noncontention approach to network access. Fiber Distributed Data Interface (FDDI) is a noncontention network topology that provides a bandwidth of 100 Mbps.

As networks have become larger, traffic has increased. One way to keep throughput up has been to segment LANs so that traffic travels only over the appropriate segment. A bridge forwards packets that don't have a local node address and directs all those that do to the appropriate node on its LAN.

Routers are more sophisticated than bridges. They're designed to support specific protocols, and also maintain several alternative paths and select the most appropriate path given certain defined conditions, such as traffic congestion.

LANs can be categorized as workgroup, departmental, backbone, and enterprise. Workgroup LANs generally consist of a few nodes that run a common traffic-intensive application such as computer-aided design or computer-aided engineering. Departmental LANs consist of users within a department that use a common application of little interest to the rest of the company. Backbone LANs act as high-speed switching stations to move data from one LAN or LAN segment to another LAN or LAN segment. Finally, an enterprise network ties together all computing resources within a company. They often link mainframes, minicomputers, LANs, and high-performance workstations.

One way to increase bandwidth is to use a LAN switch. These devices provide nodes with noncontention bandwidth when needed. Cut-through switches forward packets without waiting for the reception of an entire packet, while store-and-forward switches read an entire packet and check for errors before forwarding the packet. Ethernet switches have been enormously popular, but recently Token Ring switches have become available. They're particularly valuable because they can replace two-port bridges and connect several Token Ring segments.

Network managers seeking high bandwidth are ideal candidates for 100-Mbps "fast" Ethernet. The 100BaseT version uses the traditional Ethernet packet format but requires data-grade, unshielded, twisted-pair wire. It's challenged by 100VG-AnyLAN, a version that doesn't require data-grade, unshielded, twisted-pair wire, but does use a nontraditional Ethernet packet format. Rather than use the traditional Ethernet carrier sense multiple access with collision detection (CSMA/CD), 100VG-AnyLAN uses a different protocol, known as demand priority access method (DPAM).

Finally, Fibre Channel is a high-backwidth technology that uses fiber-optic cabling for transmission speeds that can exceed 1 Gbps. Unfortunately, there are still very few products, such as routers, available to support this technology.

2

ATM Basics

In the previous chapter, I described how local area networks function. Asynchronous transfer mode (ATM) is a radically different high-bandwidth technology that will have an enormous impact on LANs and the ways companies do business via their networks. In this chapter, I'll examine ATM basics. You'll learn how this technology works, what it can do today, and what still needs to be developed. Because there are more than two million local area networks installed worldwide, ATM will have to coexist with the older technology. I'll explain how LAN emulation works so an ATM switch can communicate with an Ethernet or Token Ring LAN. The one topic I won't examine in this chapter is how ATM functions in a wide area network; the world of global telephone communications and ATM wide area network specifications and products are discussed in depth in chapter 4. Now it's time to see why so many network managers are so excited about something as esoteric as a high-bandwidth technology.

What is ATM?

Asynchronous transfer mode (ATM) is a switching technology that uses small, fixed-size cells. In 1988, the CCITT designated ATM as the transport method, or mode, it planned to use for future broadband ISDN services. In effect, it runs on top of highly scalable physical layer protocols such as Fibre Channel and the wide area SONET protocol. You might not be familiar with some of these terms, which are usually associated with wide area networks. In the third section of this book, I'll take a close look at wide area networks, including the various technologies associated with them, such as SONET and ISDN. The important thing to remember at this point is that ATM is a very-high-bandwidth technology that's asynchronous because cells are transmitted through a network on an "as needed" basis and not merely transmitted during specific time intervals.

ATM's approach is in sharp contrast to the time division multiplexing (TDM) technology used by the voice industry for the past few years. Time division multiplexers use a fixed time-slot approach, so that if a channel doesn't have data to submit at a specific time, the slot runs empty. In other words, the trains run on time even if there are no passengers. ATM, on the other hand, adds cars full of passengers to the engine so no car is empty. ATM is inherently more efficient because its inherent bandwidth flexibility means that only as many cells are transmitted as necessary. Many ATM advocates insist on referring to this feature with some exaggeration as "bandwidth on demand."

ATM cells are small (53 bytes) compared to variable-length LAN packets. This means that the information in the header and the payload are always in the same place. Handling the cells is a simple process because ATM doesn't require additional processors or buffers to operate. Because Ethernet packets can vary between 64 bytes to more than 1,500 bytes, and Token Ring packets can contain up to 18,000 bytes, every incoming Ethernet or Token Ring packet must be buffered (stored in memory) to ensure that it's complete and free of errors before being transmitted. In contrast, ATM cells need not be buffered because of their fixed length.

ATM is a *connection-oriented* technology, in contrast to most LAN-based protocols, which are connectionless. A connection-oriented approach means that a connection needs to be established between two end points with a signaling protocol before any data transfer can take place. Once the connection is established, ATM cells are self-routing because each cell contains fields identifying its connection to the cell to which it belongs.

There are different types of transmission, including video, voice, and data, all of which can be mixed within an ATM transmission that offers scalable bandwidth from 25 Mbps to 2.5 Gbps. This bandwidth can be used by a desktop computer, workgroup, or entire network. Because ATM doesn't reserve any specific positions within a cell or specific types of data, there's very little latency (delay) in transmission; the small, fixed size of ATM cells results in predictable throughput with short delays. Switching the fixed-sized cells means incorporating the algorithms in silicon chips and eliminating delays caused by software. Another advantage of ATM is that it's truly scalable. You can cascade several switches to form larger networks.

Who Will Adopt ATM?

It isn't hard to identify the candidates for early adopters of ATM. Recently, Computer Intelligence InfoCorp identified the 500 organizations most likely to adopt ATM technology for their network backbones. The entertainment industry is likely to be among the first group to require this technology because of its need for real-time video. Other sectors generating very heavy LAN traffic that are likely to look at ATM as an upgrade to their network backbones are large financial institutions, governmental agencies, and engineering and high-tech firms. Research universities have been involved in much of the beta testing for ATM and have a substantial need for this technology in order to link together campus LANs into an enterprise network. Many large newspapers are planning to set type on their local area networks and transmit both text and pictures throughout their organization via their networks. ATM would provide them with enough bandwidth to make this task achievable.

The ATM Forum

One reason ATM has progressed so rapidly from theory to actual product release has been the absolutely amazing cooperation among vendors as well as the prodding of large end users who needs the expanded bandwidth ATM offers. The main force propelling ATM, however, has been one organization. Formed at the Fall 1991 Interop show and now with more than 700 members, the ATM Forum has been very successful in developing standards for asynchronous transfer mode (ATM).

The ATM Forum is divided into three main groups: the Technical Committee, the End User Committee, and the Marketing, Education, and Awareness Committee. The Technical Committee contains a number of subcommittees that have been busy developing specifications for such key ATM areas as signaling, traffic, testing, network management, and service applications.

The International Telecommunications Union (ITU)

Since 1948 the International Telecommunications Union (ITU) has been producing technical, operating, and tariff issue recommendations for telecommunications. A key committee, formally known as the CCITT, is now called the Telecommunications Standardization sector (ITU-T). This is a United Nations treaty organization, and as such the U.S. is a voting member. Why should you care about this exotic-sounding organization? Because ITU-T developed the first international specifications for what was then known as B-ISDN/ATM, which formed the foundation for all subsequent work by the ATM Forum.

The American National Standards Institute (ANSI)

The American National Standards Institute (ANSI) is still another organization that functions by committees. Its T1 Committee is charged with the standardization of ISDN and ATM for the United States.

The Roles and Relationships of ATM Devices

Asynchronous transfer mode is based on the concept of two end systems (terminals) communicating via set of intermediate switches. Because it's a connection-oriented system, it functions very much like the connection-oriented system you know best: the public telephone system. A switch that wants to communicate with another ATM device must first identify the address of that device. Then the system must determine the route the network will take in order to establish the precise instructions needed by each node along the network to establish the communications link. All of this internal conversation is transparent to the end user in both cases. When you make a telephone call, you certainly never think of how the call is being routed. Just as voice information must arrive sequentially to be understood, data must arrive so it can be reassembled in sequential order with minimum delay. That's the challenge as well as the opportunity for ATM vendors.

The ATM Forum has had to develop a number of different types of ATM interfaces so all types of devices can communicate. The *user-to-network interface (UNI)* is an interface between users of ATM services and ATM network nodes. A typical example is customer premises equipment in the form of an end-user device such as a computer that communicates with a public or private ATM switch. The ATM Forum has specified the physical level of this interface (what types of cabling are acceptable, the standards for formatting data, and the set of protocols or rules that must be followed). Also defined is a *network-to-network* or *network-to-node interface (NNI)*. The NNI describes a link between two switches or between two networks. Figure 2.1 describes an enterprise network using both UNI and NNI.

Because ATM is a connection-oriented network, a connection between two end points begins when one transmits a signaling request across the UNI to the network. A device responsible for signaling then passes the signal across the network to its destination end system. If this system indicates it agrees to the connection, a virtual channel is set up across the ATM network between the two systems.

Think of a virtual connection as a connection that's real but exists only for the length of the communication. Both UNIs contain mappings so the cells can be routed correctly. Each ATM cell's virtual path identifier (VPI) and virtual channel identifier (VCI) fields indicate these mappings. I'll describe these fields in more detail shortly.

Because each ATM cell contains the VPI/VCI value, even if an end system has two more virtual networks across a UNI, ATM cells can be interleaved between the two channels and the data routed to the appropriate system on its arrival. In other words, even though subway passengers bound for different final destinations can travel together on the yellow line, some passengers have written instructions that indicate to the conductor when they should be shuttled to a green-line connection. The passengers themselves (cells) don't worry about this, but leave their fate in the hands of a higher authority.

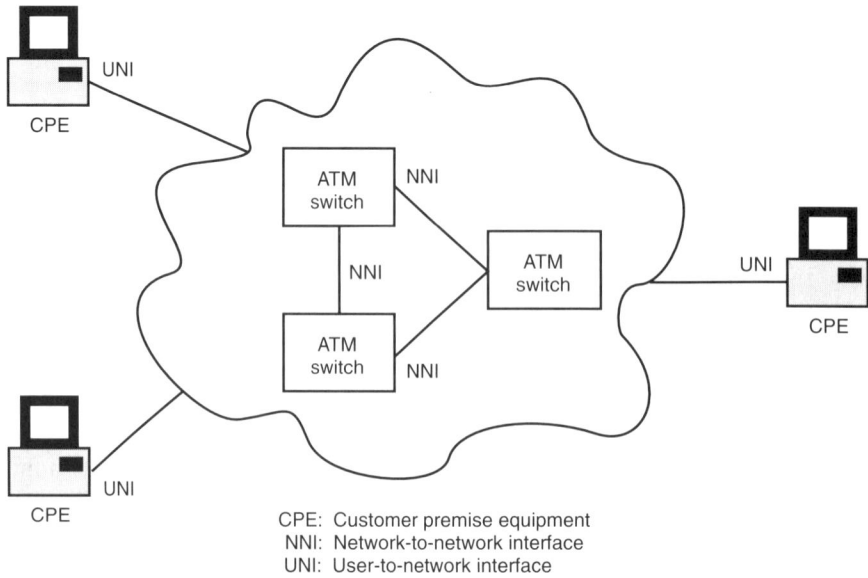

CPE: Customer premise equipment
NNI: Network-to-network interface
UNI: User-to-network interface

Figure 2.1 An ATM network and its interfaces.

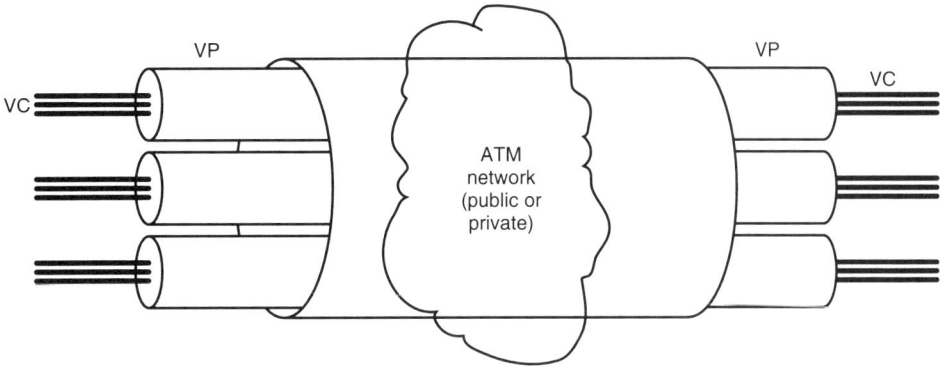

Figure 2.2 The relationship between virtual channels and virtual paths.

The ATM world is filled with switches that communicate on an "as needed" basis. Let's assume that WidgetCorp's Jim Nelson works in the company's New York City office and needs to transmit video, voice, and data to Sue Carlson in the company's Dallas office.

There are many different paths the video, voice, and data can take from the New York office to the Dallas office. Certainly, some of these paths are a lot faster than others, depending on the type of communications line used. A *virtual path (VP)* is selected for the unidirectional transport of ATM cells that belong to the virtual channels associated with a common identifier value. What this means is that Nelson's three distinctly different types of communications will have their own virtual channels, but will follow the same virtual path. Figure 2.2 shows the relationship between virtual channels and virtual paths. Note that several virtual paths with their own virtual channels can follow a common transmission path.

ATM devices that need to communicate create a virtual network, a network that stays in place for the time needed to transmit data. In much the same way, Carol Lawson might create a three-way conference call to discuss a business proposal. This "virtual network" exists only for the time the communication remains in place. When Carol completes her conversation and hangs up, the virtual network is no more. Another person could use the same lines to create another virtual network.

When I describe the specific fields within an ATM cell, some of this terminology will be easier to understand. When two ATM switches set up their call (call setup time), a *virtual channel identifier (VCI)* is assigned dynamically and this value appears in an ATM's cell header. It identifies a single virtual channel on a particular virtual path. A VC contains the instructions for sustainable cell rate; the peak cell rate that's permitted; and whether it's reserved, shared, or prioritized and with whom. It contains the maximum permitted error rate, as well as the cell delay variation. These characteristics are negotiated among all the participating pieces of ATM equipment in the path of the VC. A *virtual path identifier (VPI)* is an identifier that also appears in an ATM cell header. It identifies a bundle of one or more virtual channels (VCs).

Following Some Cells on Their Journey

As just described, there are three key bytes in an ATM cell's header that provide its network routing instructions. Two bytes are assigned to the virtual channel identifier (VCI) and one byte is assigned to the virtual path identifier (VPI). This means that there are 256 possible virtual paths and 65,536 virtual channels. Even though routing tables list the key VCI/VPI information for those who can use that particular switch, the physical switching paths are actually used when the cells are present (the virtual connection in progress). This means that two cells from different sources could have the same VCI/VPI, but the switch receiving them would know they originated from different sources and it would never be possible for them to arrive at the same time. Only one connection is possible at a time on any given virtual channel. In other words, you and your neighbor as registered telephone users could make telephone calls that follow exactly the same path and arrive at the same destination (a popular radio station offering a free car to the 18th caller), but only one of you can be the 18th caller on a particular line.

Inside an ATM Switch

Figure 2.3 shows a small ATM matrix switch. It isn't drawn to scale; it's there merely to illustrate the principles of ATM switching. ATM port cards direct the cells in a hor-

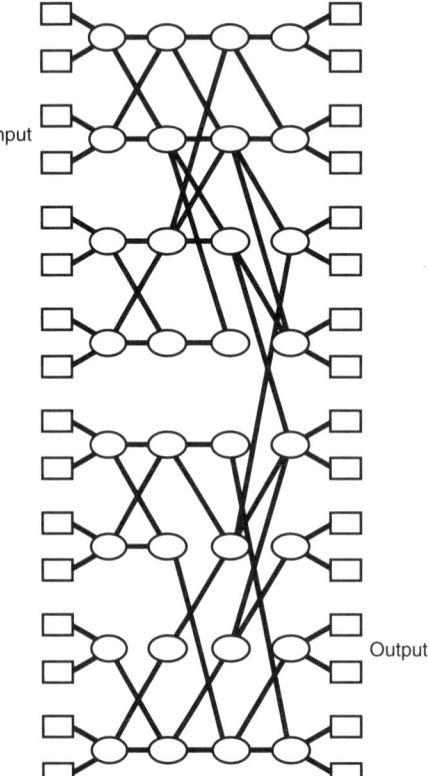

Input

Output

Figure 2.3 An ATM switch reads in and writes out cells.

izontal direction from left to right. Think of the ATM switch's matrix as a set of memory blocks. In a typical ATM switching matrix, a cell would enter horizontally and its contents would be written into these memory blocks as soon as it arrives. Each memory block can hold the contents of several cells. Each cell's VPI/VCI, located in its header, is checked. If the cell doesn't contain a VPI/VCI, it's discarded. If it is, then its VPI/VCI is checked against a routing table to determine the new required VPI/VCI. If the switch contained a 16×16 matrix, then memories of the cell's contents are erased from the other 15 memories associated with the horizontal path it entered. The remaining cell, now with an updated VPI/VCI, must compete with other cells in its vertical path for transmission. When it's read out by the switch depends on the priority it has been assigned as well as several other factors.

The ATM Protocol Reference Model

Just as the OSI layered protocol model describes communication between two computers over a network, the ATM protocol model describes how two end systems communicate via ATM switches. As shown in Figure 2.4, the key layers that need explanation are the ATM Adaptation, ATM, and Physical layers. Let's stop for a moment and follow data as it travels over an ATM network using the protocols (or rules) associated with the ATM model.

The ATM Adaptation layer (AAL). A user transmits a data packet designed for another user. The ATM adaption layer (AAL) provides services to the higher layers that support classes of service for transported data. Its major concern, though, is the segmentation and reassembly of data. It takes this data and splits it up to multiple 48-byte cells.

The ATM layer. The ATM layer is responsible for providing the appropriate routing information for cells in the form of VPI/VCI values, which are part of the control information found in a cell's five-byte header. The VPI/VCI values (local to a specific switch) ensure that the cell will exit the correct switch output port. The ATM layer is also responsible for ensuring that cells are ordered (stay in the correct order). This layer passes control over the cells to the physical layer at a switch's output port.

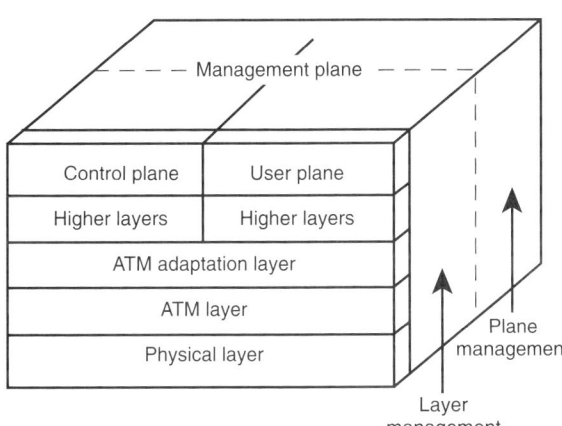

Figure 2.4 The ATM protocol reference model.

The physical layer. The physical layer is primarily responsible for transmitting and receiving data. Then it groups these cells in payload envelopes, adds routing information, and transmits them. The process is reversed at the other end of the network when the destination node begins the process of translating the cells back into data.

The part of the layered architecture used for end-to-end or user-to-user data transfer is known as the *user plane (U plane)*. The control plane defines higher-level protocols used to support ATM signaling, and the *management plane (M plane)* provides control of an ATM node and consists of two parts: plane management and layer management. The plane management function manages all other planes and the layer management function is responsible for managing each of the ATM layers.

The physical layer

The physical layer defines the physical interfaces and framing protocols associated with ATM. This layer is segmented into two sublayers: the *transmission convergence (TC) sublayer* and the *physical medium dependent (PMD) sublayer*. The reason for sublayers in this ATM architecture is to decouple the transmission from the physical medium to permit a variety of physical media.

The TC concerns itself with adaption to the transmission system, defined as the reception of cells from the ATM layer and their packaging into appropriate format for transmission over the PMD. The TC also handles cell delineation, cell scrambling/descrambling, and cell-rate decoupling. Cell delineation is the extraction of cells from the bit stream received from the PMD. The function of cell-rate decoupling is to insert/suppress idle cells to or from the payload in order to provide a continuous flow of cells. Finally, the TC generates and verifies the *header error check (HEC)*. It calculates the HEC from the bits received and checks it against the HEC value of the received cell. If there's a match on consecutive cells, then the TC assumes correct cell boundaries. If there's no match for many successive cells, the TC knows that the correct cell delineation isn't yet found.

The ATM layer and ATM cells

The ATM layer performs four basic functions. It multiplexes and demultiplexes cells of different connections. Multiplexing refers to the process of taking several different data streams and consolidating them into a fast-flow data stream. At the other end of the communication path, demultiplexing reverses the process and directs the data back to its appropriate data stream and toward its ultimate destination. These connections are identified by their virtual channel identifier (VCI) and virtual path identifier (VPI) values. The ATM layer also translates VCI and/or VPI values at the switches or cross connections, if required. It's responsible for extracting/inserting the header before or after the cell is delivered to or from the ATM adaptation layer. Finally, this layer implements a flow control mechanism at the universal network interface (UNI) by using the general flow control (GFC) bits of the header.

Figure 2.5 describes the ATM cell format, which consists of a five-byte header and a 48-byte information field. The cell size is deliberately small so there's very little delay or latency in the delivery of information, particularly real-time video data.

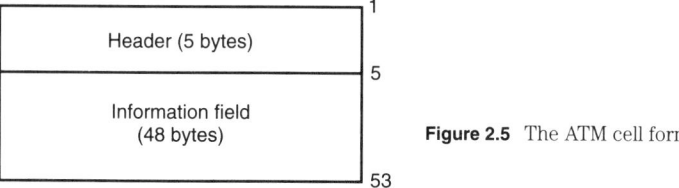

Figure 2.5 The ATM cell format.

Across the UNI

Generic flow control		Virtual path identifier		1	
Virtual path identifier		Virtual channel identifier		2	
Virtual channel identifier				3	Bytes
Virtual channel identifier	Payload type indicator		Cell loss priority	4	
Header error control				5	

Across the NNI

Virtual path identifier				1	
Virtual path identifier		Virtual channel identifier		2	
Virtual channel identifier				3	Bytes
Virtual channel identifier	Payload type indicator		Cell loss priority	4	
Header error control				5	

Figure 2.6 ATM cell header formats across the UNI and NNI.

The ATM cell header

Figure 2.6 describes the ATM cell header fields found in cells going across the UNI and NNI. The four-bit *generic flow control (GFC)* field is used only across the UNI to control traffic flow and prevent overload conditions. This field isn't defined across the NNI, and the corresponding bits are used for an expanded virtual path identifier (VPI) field.

Two of the fields in the cell header have already been discussed in this chapter. The virtual path identifier (VPI) field is used to identify virtual paths. Consisting of eight bits across the UNI and 12 bits across the NNI, the field isn't defined by either the CCITT or ATM Forums. The virtual channel identifier (VCI) field is 16 bits long. End devices assign a value to the VPI and VCI fields when requesting a connection to an end system.

The *payload type identification (PTI)* field consists of three bits and is used to identify the payload type carried in the cell, as well as to identify control procedures. The ATM Forum designates the setting of one bit to indicate congestion, a second bit for network management, and a third bit to indicate an error condition.

The *cell loss priority (CLP)* field is a single bit that indicates a cell's loss of priority. This bit is set to 1 when a cell can be discarded due to congestion; if a switch experiences congestion, it will drop cells with this bit set. This results in giving priority to certain types of cells carrying certain types of traffic, such as video in congested networks.

The *header error check (HEC)* is an eight-bit cyclic redundancy code that's calculated over all fields in the ATM header. This type of error checking can determine all single-bit errors and a number of multiple-bit errors. Error checking is very important in ATM operations because an error in the VPI/VCI could corrupt the data flow of their circuits.

General ATM Operation

ATM requires that a connection be made between two end points before information can be exchanged. An end point on a network sends a signal across the UNI to the network requesting a connection to another point. The network sends this request to its destination point, where it's interpreted. If this node accepts the request for a connection, a virtual channel is established across the network. Two end or switching points can be linked via a virtual channel link. The VPI and VCI fields of the ATM cell header contain the routing information that's required.

Figure 2.7 illustrates how this process works. End point A requests connection to point B. The ATM network assigns value P to the VCI of A, and value Q to point B. Node A will use P for outgoing information, and B will use Q for incoming information. Lookup tables are set up throughout the network. The same process is followed for the reverse direction. The first switching node that receives a cell from A will consult its lookup table to find out where the cell should be switched to and what value the outgoing VCI should be assigned to. This process is repeated until it arrives at B. I'll return to this process a bit later in the chapter and use some concrete numbers to shed even more light on this crucial task.

How does the ATM layer function when the ATM node is an end system? The AAL layer provides it with information. When the ATM layer exchanges a cell stream with the physical layer, it inserts this information as well as the required parameters in the header fields, including the crucial VPI/VCI values. If it has no information to transmit, it fills the information field with idle cells. the ATM layer is also responsible for controlling the quality of service for each circuit, a value that's negotiated when

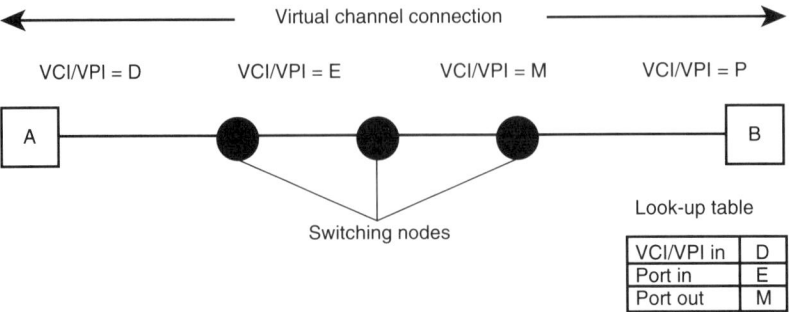

Figure 2.7 An ATM virtual channel connection.

circuits are established. Among the parameters that are negotiated are the peak and average data rates, the acceptable delay, and the loss rate.

The ATM layer's operation is even less complicated for a switch. the ATM layer under this circumstance receives an ATM cell on one port and uses the VPI/VCI value to determine to which port to forward the cell. It then forwards the cell to the appropriate port, changes the VPI/VCI to reflect the cell's routing, and transmits the cell to the physical layer of that port.

The Four Classes of ATM Traffic

A major advantage in an enterprise network environment is its ability to handle a variety of different types of traffic. Current specifications define four different classes of traffic that can be handled. ATM is designed to handle the different classes of traffic listed in Table 2.1.

Let's take a closer look at how these levels of service are handled by the ATM adaption layer (AAL) protocols. The AAL provides service to the higher layers that correspond to the four classes of traffic. Figure 2.8 describes the service classes associated with the AAL protocols, while Table 2.2 describes the protocols associated with these services.

It's important to distinguish between the service offered to the higher layers by the AAL and the type of traffic carried by the AAL protocol across the network. Type 1 protocol is designed to handle the needs of *constant bit rate traffic (CBR)* network services. An example of this type of traffic is a T-1 (1.544-Mbps) signal that's used to transfer bits at a constant rate between two points. While AAL type 1 typically offers class A service and carries class A traffic, the type of service and the type of traffic carried by the AAL don't always have to correspond. Class A is associated with connection-oriented, constant-bit traffic. A higher layer provides the AAL with

TABLE 2.1 Classes of ATM Traffic

Class of service	Old class of service	Description
CBR	Class A	Constant bit rate (CBR), connection-oriented, synchronous traffic (uncompressed voice or video); supports peak cell rate traffic
VBR-RT	Class B	Variable bit rate (VBR) real-time, connection-oriented, synchronous traffic (live video transmissions); supports peak cell-rate traffic, sustained cell-rate traffic, and maximum burst-size traffic
VBR-NRT	Class B	Variable bit rate non-real-time traffic (video playback, multimedia); supports peak cell-rate traffic, sustained cell-rate traffic, and maximum burst-size traffic
ABR	Class C	Variable bit rate, connection-oriented, asynchronous traffic (wide area X.25, frame relay over ATM); supports peak cell-rate traffic and maximum burst-size traffic
UBR	Class D	Connectionless packet data (LAN traffic, wide area SMDS traffic, etc.)

	Class A	Class B	Class C	Class D
Timing and relation between source and destination	Required		Not required	
Bit rate	Constant		Variable	
Connection mode	Connection-oriented		Connectionless	
AAL types	1	2	3/4,5	3/4

Figure 2.8 AAL layers.

TABLE 2.2 Quality of ATM Service

Service class	Congestion feedback	Bandwidth guarantee	Throughput guarantee
CBR		√	√
VBR		√	√
UBR			
ABR	√		√

a fixed number of bits at a regular frame rate. The AAL delivers the bits to their destination at the same bit rate. Generally the ATM layer uses priorities to ensure that the ATM cells are sent and received at the same rate across an ATM network.

Under CBR, a network application establishes a CBR connection and negotiates what's known as a *peak cell rate (PCR)*, the maximum data rate the connection will support without losing data. Traffic is then transmitted at that rate. The ideal type of traffic for CBR service is real-time voice and video traffic since both require constant data streams and cannot tolerate lost data.

Class B traffic is connection-oriented, but permits a variable bit rate. Data is passed to the AAL from the higher layers at fixed intervals, but the amount of data might vary from transmission to transmission. If the amount of data exceeds the capacity of a single cell, then the data is segmented into multiple cells and reassembled at the data's destination.

Variable bit rate (VBR) service requires negotiation not only for the PCR but also the *sustained cell rate (SCR)*, which refers to the average throughput rate the application is permitted. Part of the VBR negotiation actually consists of determining how long transmission will stay at the PCR rate (known as the *burst tolerance*). In other words, under VBR the traffic can soar above SCR to PCR for short periods, but the VBR connection will maintain the SCR as the average rate by dropping traffic flow to a lower rate for the time needed to achieve the SCR. VBR users have a guaranteed quality of service regarding cell loss and bandwidth availability as long as the traffic meets the criteria negotiated.

VBR is designed for transaction processing and LAN-to-LAN traffic because of their bursty nature. Unfortunately, LAN traffic is especially bursty and can exceed any negotiated parameters. Until recently, network managers had a number of choices, all of which were rather bleak. They could order more bandwidth than they needed most of the time (the risk-averter approach), try to estimate carefully (the bean-counter approach), or they could just assume that bits would be lost occasionally (the Russian Roulette approach). Now there's another approach available: the available bit rate (ABR).

Available bit rate (ABR) addresses many of the concerns that have troubled network managers about VBR and provides reliable delivery of bursty traffic. It uses excess bandwidth and network management algorithms to evaluate network congestion and eliminate cell loss. ABR negotiations include the PCR I've already discussed as well as a minimum cell rate (MCR). The MCR (cells/second) relates to an application's ability to handle latency. ABR provides a guaranteed quality of service concerning bandwidth availability and cell loss. It does not guarantee against cell delay, so non-real-time LAN applications are ideal for ABR.

Unspecified bit rate (UBR) lends new meaning to the term *service*. It has no specified bit rate and no quality-of-service guarantees. In fact, the only assurance is that UBR will make its best effort to deliver cells. There's no flow control, so if traffic becomes very heavy then cells will be lost once buffers are full. Think of this service as what you'd expect when you send a letter to someone in Romania. It might get there, but then again it might not.

Class C traffic is connection-oriented, variable-rate data with no timing relationship between source and destination. Packet-switched networks such as X.25 and frame relay require this type of service. This is a very key part of the AAL because it will probably be the heart of a wide area network transmission. The problem is that it's a very complex, layered protocol and difficult to define. Much of it is based on the IEEE 802.6 protocol for metropolitan area networks. There's now a complete AAL standard for class C, type 5: the simple subset of AAL 3/4.

The AAL type 3/4 protocol has been developed to support the carrier's switched multimegabit data services (SMDS) connectionless service, which I'll describe in chapter 4 when discussing wide area networks. It's compatible with the IEEE 802.6 protocols for metropolitan area networks that support SMDS. This is a bit tricky because SMDS uses variable-size frames that must be delivered error-free. The AAL type 3/4 protocol can handle this problem by using a transmitter to segment each SMDS frame into fixed-size segments and then placing them into ATM cells. A receiver is also needed to reassemble the SMDS segments into the original frame. The entire process is known as *segmentation and reassembly*.

Flow Control on ATM

The ATM Forum has adopted a rate-based approach to flow control known as the *proportional rate control algorithm*. This approach allows transmitting stations on an ATM network to increase or decrease their transmission rates according to the network's available bandwidth, reducing congestion and cell loss. Transmitting stations collect bandwidth information by generating a resource management cell for every so many cells that are transmitted. The resource management cell collects bandwidth information along its way to its destination node, which returns the resource management cell and bandwidth data to its transmitting source so subsequent transmissions can be optimized.

LAN Emulation

It's going to be a long time before there are "real" ATM networks. So many of the features that make ATM so desirable, such as its quality of service parameters, are designed for applications that are specifically designed to run in conjunction with ATM. For the next few years, the majority of applications running on LANs will take advantage of ATM's high bandwidth and that's about it. Most LAN applications will be fooled by the ATM Forum's LAN emulation protocol (described in detail in this section) into thinking that they're communicating with a traditional (legacy) LAN. Current driver interface specifications, such as Novell's open data-link interface (ODI) and Microsoft's network device interface specifications (NDIS), won't change. The LAN emulation software will ask the ATM switch to provide its best-effort service using an unspecified quality of service.

Eventually, "native" ATM applications will be written to take full advantage of ATM's features. In addition to changes in driver interfaces, this will also require major changes in transport layer protocols and desktop operating systems. Novell and Microsoft have already indicated that they will make these changes. Figure 2.9 shows how a native ATM application will work over a LAN.

Novell has announced a strategic alliance with Efficient Networks that will enable it to build ATM capabilities into NetWare. Such support would enable applications for NetWare LANs to use ATM's massive bandwidth. Today, there's LAN emulation support via NetWare ATM Ethernet LAN emulation.

Microsoft is planning to add extensions to Windows NT's network driver interface specification (NDIS) so users will have out-of-the-box support for native ATM appli-

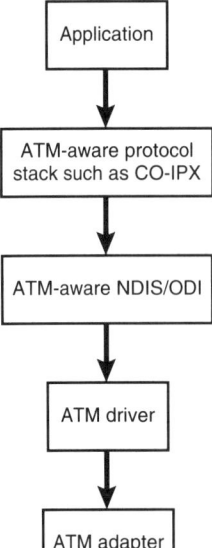

Figure 2.9 How a "native" ATM application transmits data over a LAN.

cations, LAN emulation, and IP over ATM. Microsoft has licensed Olicom's ATM technology to help it achieve this goal.

Microsoft has announced to developers that it plans to release its ATM software in the Cairo release of NT, scheduled for late 1996 or early 1997. The major advantage of incorporating ATM support in the NOS rather than on an adapter card is that it will probably be a bit faster and more efficient. Microsoft will use its extended NDIS interface in conjunction with the ATM Forum's user-to-network interface (UNI) call management software to provide a native ATM interface. Microsoft has also indicated that it won't provide an ATM LAN emulation server but will provide an ATM LAN emulation client. It also will add IP over ATM to its NT TCP/IP stack with an ATM module that resolves IP addresses to ATM addresses.

The first generation of ATM products were generally used in self-contained workgroups consisting of a few high-performance workstations. They weren't part of local area networks because there was one crucial piece missing from the set of ATM specifications developed by the ATM Forum. The missing link in this case was LAN emulation.

With millions of local area networks already installed, clearly network managers couldn't change their LANs to accommodate ATM. It was up to the ATM Forum to develop a method of LAN emulation that would make LAN-to-ATM communications virtually seamless as well as transparent to end users. It wasn't a trivial assignment. The two technologies could not be more different.

As discussed earlier in this chapter, LANs are connectionless. When a node transmits a packet of information onto a LAN, it's a shared environment and not a private connection between sender and receiver. ATM of course creates a virtual connection for the time of the transmission. LANs use variable sized packets that must be buffered until received completely and then retransmitted, while ATM cells are uniform in size and don't need buffering nor even extra processing power.

Vendors didn't wait for the ATM Forum to complete its LAN emulation specifications. Many, such as Bay Networks, jumped into the market with makeshift, kludgy (inelegant) boxes that stood between LANs and ATM switches to perform the necessary conversion. The raw LAN packets went in one side, and processed sausage in the form of ATM cells came out the other side. These products were clearly interim solutions until the ATM Forum completed its deliberations.

Today there's a set of ATM LAN emulation specifications. While some details might change by the time this book is printed, the basic structure won't change, and that's the thrust of this chapter. LAN emulation is designed to function at level 2 of the OSI model, which means that it's independent of upper-level protocols. That also means it doesn't really care if the LAN data is coming from a NetWare environment with the IPX protocol or a UNIX environment with the TCP/IP protocol. The ATM switch doesn't concern itself with LAN emulation, but rather with establishing a virtual connection and actually switching the cells. The emulation resides elsewhere, as you'll see shortly.

The ATM Forum has developed the *LAN emulation user-to-network interface (LUNI)*, pronounced "loony." In effect, this protocol fools a LAN into thinking it's communicating with an Ethernet or Token Ring LAN. Emulated LANs can emulate only one topology (Ethernet or Token Ring) at a time. If an FDDI LAN needs to be incorporated into an ATM backbone, then its packets must first be converted to Ethernet and then bridged so only one topology is being emulated at a time.

The major components of ATM emulation are the *LAN emulation client (LEC)*, the *LAN emulation server (LES)*, and the *broadcast/unknown server (BUS)*. All devices that communicate with devices attached to traditional LANs (servers, routers, etc.), qualify as LECs. Venders can put LEC software on a variety of different devices, including workstations and servers integrated as device drivers. They can also integrate this software into routers and switches. Each emulated LAN can have a maximum of 65,278 LECs. Figure 2.10 shows that the LAN emulation client software resides at the data link layer of the OSI model.

The LES software can reside on any ATM device on a LAN and is responsible for resolving MAC and ATM address mappings for all emulation LAN clients. The BUS transmits broadcasts as well as multicast packets to all emulated LAN clients. If the LES cannot map a destination address for a data packet, it's broadcast to all clients on the LAN. Let's spend a few moments looking at the LAN emulation process in more detail. Assume you can actually watch the events taking place on an ATM adapter with LAN emulation software operating.

Also assume that you have a multimedia server with some data that it wants to send to an Ethernet LAN client. First it must communicate with the LES in order to learn the ATM address of an LEC for which it knows the MAC address. The LES acts as a clearing house for address resolution requests. The protocol used by an end station or a bridge to resolve a LAN address to an ATM address is known as the *LAN emulation address resolution protocol*. In order to establish this communication with the LES, the LEC must set up a *virtual channel connection (VCC)*. Once the connection is established, the two exchange information. This first communication between the two serves as the LEC's registration. The LES adds this information to its table so that other LECs can learn its address. It also provides the LEC with a two-byte identity code that's unique for the LEC on the emulated LAN. The LEC uses this

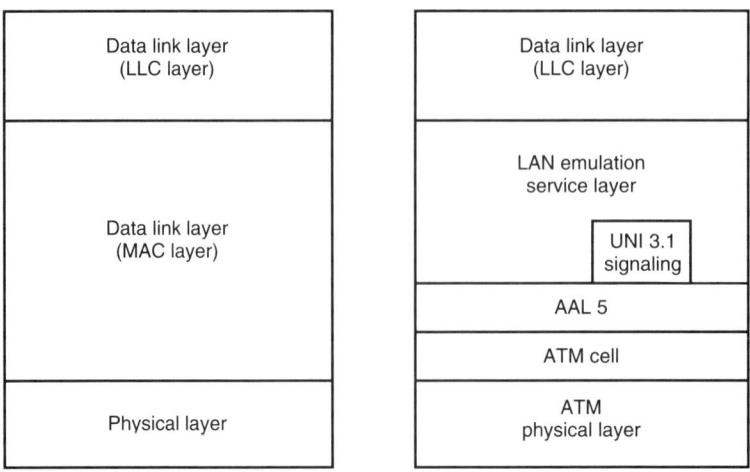

Figure 2.10 The ATM emulation software and the OSI model.

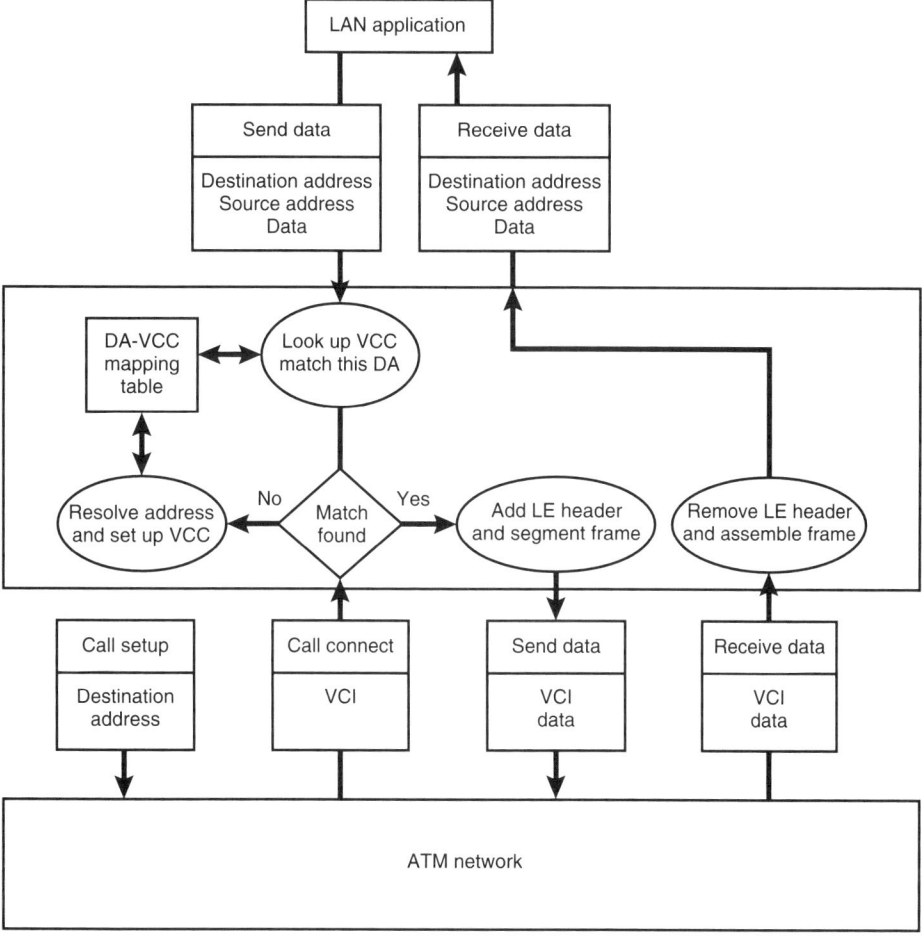

Figure 2.11 The LAN emulation process. *(Madge Networks)*

identity code as its LAN emulation header that it appends to all Ethernet or Token Ring frames before segmenting them into cells.

If the LES doesn't know the MAC address that has been requested by a specific LEC, it forwards the frame to the BUS via a VCC that it establishes. The BUS treats this frame as if it were a broadcast or multicast and forwards it to all registered LECs. Figure 2.11 shows the process I've just described. Figure 2.12 takes the process a bit further and shows how an Ethernet or Token Ring packet goes through the LAN emulation layer and is converted into ATM cells.

The LAN Emulation Configuration Server

When I explained how LAN emulation works, I assumed that the LEC knew the address for its LES in order to communicate with it. What if it doesn't have this infor-

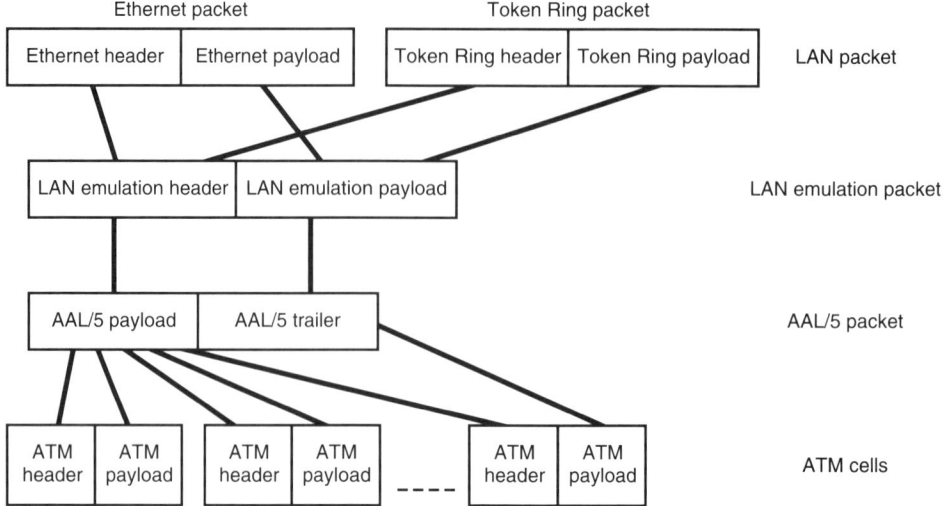

Figure 2.12 LAN emulation converts packets into cells.

mation? Another possibility is that there might be several LESs on an ATM network. Which one should it contact? LAN emulation enables network managers to define a *LAN emulation configuration server (LECS)*. An LEC can query an LECS for the address of the appropriate LES with which it needs to register.

A virtual channel is established between the LEC and LECS, and then the LEC issues a configuration request. It tells the LECS the LAN topology it's emulating (Ethernet or Token Ring), the maximum packet size supported, the emulated MAC address of the LEC, and the name of the emulated LAN with which it wants to communicate. The LECS then tells the LEC the ATM address of the appropriate LES.

Internet Protocol Over ATM

While the ATM Forum's LAN emulation specification will carry the bulk of network traffic over ATM, another approach is available. The Internet Engineering Task Force (IETF) has defined an IP-specific media access control (MAC) layer for ATM: *classic IP over ATM*, found in request for comment 1577.

This specification supports only one protocol—IP—and doesn't emulate the existing MAC layer; therefore, it's a much simpler and more limited approach. Any device with a legacy LAN adapter (such as Ethernet or Token Ring) must use a router or bridge to access a node or server using classic IP over ATM.

Because the IETF didn't have accommodate the existing LAN protocols, it could optimize its protocol. Its approach uses less overhead because it allows a very large frame with a maximum size of more than 9,000 bytes. Finally, users can invoke various ATM features, such as quality of service, maximum cell delay, or acceptable cell loss, when using this protocol. Some vendors, including IBM, Efficient Systems, and Fore Systems, ship products that support RFC 1577.

ATM Network Management

There are two primary functions associated with network management of an ATM network: internal connection management and external network management. Internal connection management is responsible for call setup, call routing, address resolution, and management of switched virtual circuits and permanent virtual circuits. The internal network management function must be contained within the ATM switch and ATM adapters because call setup must be performed continually and as quickly as possible by the ATM network.

Connection management is also responsible for performing automatic network configuration, which permits discovery of end nodes, switches, and other ATM-attached devices. This approach eliminates the trouble and expense associated with moves and changes in traditional LANs. The connection management software also monitors performance of every port and every virtual path, trunk, and channel, and collects statistics on usage.

The ITU-T recommendation I.371 specifies how a traffic contract defines user traffic management. For every virtual path or virtual channel connection, there's a contract between the user and the network regarding the quality of service to be provided. Traffic can include a sustained cell rate (SCR), a minimum cell rate (MCR), a peak cell rate (PCR), and a burst tolerance (BT).

Much work still needs to be done. At this time, for example, there's no ATM equivalent of status signaling for use by the internet protocol's address resolution protocol (ARP). This means that there's no way an IP user can check a permanent virtual circuit's status, or a network can inform a user of this circuit's status change. The real significance of this limitation is that, when conditions change, a router can't determine what is connected at the other end of the permanent virtual circuit.

While terminal emulation is one of the major issues associated with ATM on local area networks, an equally serious issue is the fact that LAN protocols aren't designed to run efficiently with ATM on LANs. In addition to working on ATM emulation, as noted earlier in this chapter, Novell has announced that it plans to develop a more efficient version of its internetwork packet exchange (IPX), which will offer native ATM transport services. IPX was designed originally for broadcast, connectionless networks; this means that it doesn't support the virtual connections used in ATM's connection-oriented technology.

A second problem with the current version of NetWare is that its servers use Novell's service advertising protocol (SAP), which periodically broadcasts the availability of its services across a local area network. In a connectionless environment, clients simply pull the message off the network, but that's not the case in a connection-oriented network where these periodic broadcasts would tie up the limited number of virtual circuits supported by workgroup switches.

External Network Management of ATM Networks

While several ATM vendors have developed their own proprietary network management software, the real future will be in SNMP-compliant software. The first stage, now underway, is for ATM vendors to provide SNMP MIBs for the various elements within their products. These MIBs permit network managers to use standard SNMP management systems to monitor and configure ATM networking equipment. The MIBs are likely to show up first as vendor-specific MIBs. A second stage in ATM network management development will be standard MIBs with vendor-specific extensions. Vendors at that point will then provide ATM network management software that will run on existing management platforms but offer even more advanced capabilities.

The Internet Engineering Task Force (IETF) is now working to define an ATM management information base in conjunction with the ATM Forum. At the same time, the International Telecommunication Union's Telecommunications Standardization Sector and the American National Standards Institute (ANSI) are working to develop network management specifications for public networks. When the work of these two very different groups eventually converges, SNMP will most likely manage private ATM networks while the common management information protocol (CMIP) will be used within and between public networks.

Figure 2.13 describes the flow of management information in an ATM network. As an example, M1 could represent a Token Ring interface while M2 is SNMP over IP. M3 is also SNMP over IP, while M4 shows SNMP or CMIP over X.25 or IP. Finally, M5 is the CMIP interface between public networks.

The ATM Monitoring Group and Its Goals

There are several network management problems unique to ATM. The sheer speed of this technology creates management problems. Buffers can overflow in a wink of an eye and thousands of cells can be dropped. The hardware-centric nature of ATM creates problems for the hardware designed to monitor this technology. Perhaps ATM does it job too well. Switches capture header information and then throw away cells containing errors before test equipment can analyze them. And if that isn't a tough enough challenge, diagnostic equipment must also be ca-

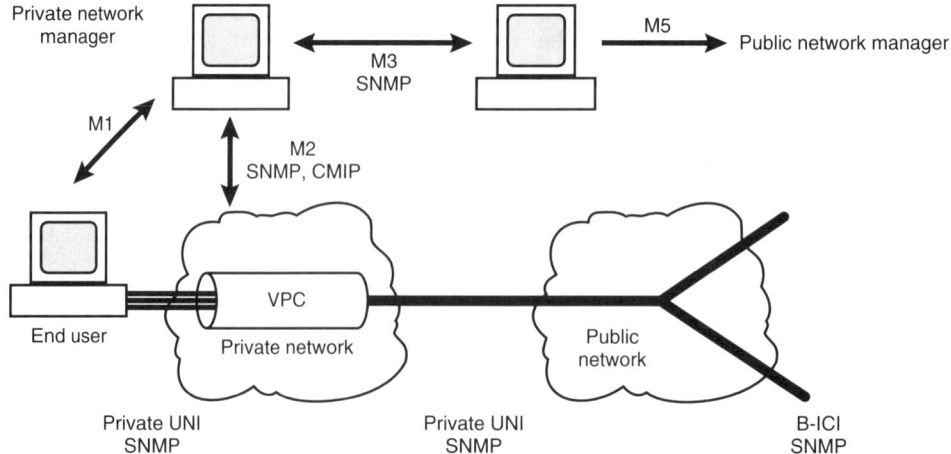

Figure 2.13 Management information flow in an ATM network.

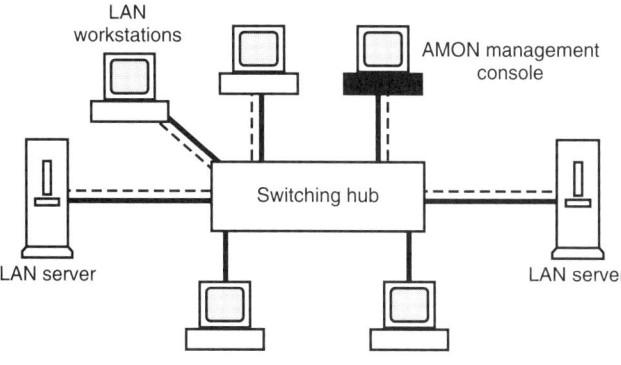

---- Virtual circuit

Figure 2.14 AMON management: A virtual circuit between two nodes is copied, and that information is sent to a stand-alone management console

pable of handling legacy LAN protocols, LAN emulation protocols, IP over ATM, and ATM protocols themselves.

Several of the leading ATM switch, remote monitoring (RMON), and test equipment manufacturers established the ATM Monitoring (AMON) group in 1995. It plans to work closely with the ATM Forum. The purpose of this group is to establish an open standard for an ATM circuit steering management information base (MIB) so the contents of virtual circuits can be directed to testing and monitoring equipment.

While there are already ATM MIBs, such as ATM MIB, SONET MIB, and DS-3 MIB, that monitor the lower levels, such as the physical and ATM layers, the group believes that network managers need to be able to monitor the upper levels (3 and above) to resolve protocol problems and analyze network traffic. The result will be the *ATM circuit steering management information base (ACS MIB)*, a database that will enable analysis tools to work in an ATM environment.

ATM provides a mechanism for creating copies of virtual circuit traffic in order to multicast them to specific locations within an ATM network. The ACS MIB will take advantage of this multicasting capability to direct the traffic to test equipment for decode analysis. This will enable network managers to look at the actual traffic traveling over the network.

It appears that this MIB residing on an ATM switch will be compliant with the simple network management protocol (SNMP), version 1, rather than version 2, although it will be readable from any version-2-based system. The monitoring devices produced to read the ACS MIB will require an ATM adapter to connect to the network. The first products are likely to require one management console for each ATM switch to be managed. Eventually, though, multiple switches are likely to be manageable from a single console.

Let's consider an environment with an ATM switching hub that requires management via AMON. It's possible to copy a virtual circuit between two nodes and send that information to a stand-alone management console. The ATM monitoring group's MIB then lets the network manager view the circuit without having to connect directly to the nodes. Figure 2.14 illustrates this procedure.

Security

ATM doesn't currently have any hooks with which to handle security. An ATM Forum subcommittee has been working on defining security specifications to address issues such as access control, auditing, authentication, confidentiality, data integrity, and encryption key management.

Until now, customers have been using ATM in conjunction with the transmission control protocol/internet protocol (TCP/IP) in order to use TCP/IP's security, but this approach doesn't work with native ATM. Because ATM is currently used for the most part internally within most companies, security hasn't yet become a major issue. All this will change, however, when ATM is transmitted over wide area networks.

ATM Private Network-to-Network Interface

Network managers today are reluctant to mix and match ATM switches. A *private network-to-network interface (PNNI)* specification will make it more likely that ATM switches from different vendors will interoperate.

In order for PNNI to permit this level of interoperability, it must deal with the tricky issue of routing. An ATM network can contain thousands of connections, some at constant bit rates, other at variable bit rates, and still others at available bit rates. The routing divisions for all these connections must ensure quality of service to a wide variety of traffic types.

PNNI is a hierarchical, link-state, routing protocol. What this means is that its purpose is to distribute topology information between switches so paths can be calculated between network end points. To do so, it must also be able to extend the user-to-network interface signaling protocol so capabilities such as alternate routing on link failure are permitted. If PNNI works, then it won't be necessary for each ATM switch to keep detailed pictures of the total network topology and knowledge of each node on the network.

PNNI uses the concepts of clusters of switching systems. In each peer group, or cluster, the nodes exchange information so they have identical views of the network. Nodes query each other to see if they're in the same cluster. The answer is yes if they have the same identification numbers. Peer nodes have complete information on each other and only summary information on other network nodes. Border nodes have information on at least one link that crosses their cluster boundary.

Every time a connection request is received, the switch computes a path based on the node's topology database in conjunction with traffic characteristics and quality-of-service requirements. The *quality of service (QOS)* includes such variables as cell loss ratio, cell delay, and delay variance. An ATM network must negotiate these parameters, depending on the different types of traffic and the traffic requirements for applications such as voice, data, and video. The network must establish traffic prioritization and congestion management. The PNNI protocol finds a path across an entire multivendor network of switches that's consistent with the quality of service requested. If a connection setup fails, then PNNI partially releases that setup in progress and the connection request is rolled back to an alternate switch.

Plans call for PNNI phase 0, known as *interim interswitch signaling protocol (IISP)* to provide a rudimentary level of interoperability among different vendors' switches. One of the major limitations of this first phase of PNNI is that it comes with the overhead of manually configured static routing.

As plans continue for phase I of PNNI to be implemented around the third quarter of 1996, there's also a push to extend PNNI's capabilities to networks that comprise ATM switching systems and existing routers. This movement is known as *integrated PNNI (IPNNI)*, a protocol that would permit legacy routers to be part of the PNNI peer nodes. Eventually IPNNI could replace frame-only protocols, such as Cisco's IGRP.

Congestion Management

Congestion management is a major problem for ATM because of the need to enforce network parameters, such as committed information rate (CIR), cell loss ratios, delay and delay variance, and burst size and duration—all of which must be managed in a nonblocking environment.

Switched Virtual Circuits

ATM permits both permanent virtual circuits (PVCs) and switched virtual circuits (SVCs). The advantage of SVCs are that this feature enables a switch to dynamically establish connections. PVCs must be set up in advance. The ATM has adopted the Q.2931 specification for setting up SVCs. From a customer's perspective, it's much better to purchase a switch that supports this specification than a proprietary protocol. PVCs are used when only a few connections are needed. Because PVCs are defined in terms of their end points, a network could vary with each exchange. Devices linked by a PVC must maintain tables that track all connections. For this reason, workstations joined by a PVC would require tables listing every other workstation on the network. This is clearly not practical and helps explain the purpose of an SVC.

To make matters even more complex, SVCs must be able to dynamically negotiate each of the ATM quality-of-service parameters with the network in real time. Congestion management is responsible for seeing that each switched virtual circuit performs up to the conditions established in the QOS parameters.

One problem that currently exists with SVCs is interoperability. A large part of this is how long an SVC's bandwidth remains committed when there's no traffic before the network takes down the connection automatically so the bandwidth can be dynamically reallocated. Current SVC standards don't specify an inactivity time-out. Currently vendors are building different values for inactivity time-outs into the driver software associated with their ATM network interface cards.

The User-to-Network Interface (UNI version 3.1)

The Signaling Working Group of the ATM Forum's Technical Committee has issued version 3.1 of the user-to-network (UNI) recommendation. It explains the requirements for setting up and tearing down switched virtual circuits and permanent virtual circuits. Addresses use the same format defined in layer 3 of the OSI model. This UNI version also provides specifications for using simple network management protocol (SNMP) for local connection management services across the UNI. It also defines an ATM UNI management information base (MIB) known as the *interim local management interface (ILMI)*, which uses SNMP to transmit local signaling management data across the UNI.

Version 3.1 is in conformity with the Q.2931 signaling specification of the International Telecommunications Union's Telecommunications Standards Sector (the old CCITT).

ATM's Physical Interfaces

Earlier in this chapter I pointed out that ATM was originally designed as a transport layer that would run on top of such physical standards as SONET. In the mid-1980s, Bellcore proposed a number of physical layers as a standard for optical fiber transmission lines for the U.S. public telephone system. *Synchronous optical network (SONET)* was designed to link transmission media from different vendors. In Europe, SONET because known as the *synchronous digital hierarchy (SDH)*. Table 2.3 describes the different SONET specifications.

There are over 13,000 installed *transparent asynchronous transmitter/receiver interfaces (TAXIs)*. Running at either 100 Mbps or 140 Mbps, the vast majority are found in campus environments and use multimode optical fiber. These have been ignored by the ATM Forum, which has developed a 155-Mbps specification for SONET framing (OC-3) running on category 5, unshielded, twisted-pair wire. It has also developed a specification for 51.84 Mbps (SONET OC-3) over category 3 UTP. The 622-Mbps specification also uses the SONET OC-12 standard and is expected to grow in popularity. Another specification expected to grow in popularity is ATM over North American T-1 (1.544 Mbps). The following is a list of the various ATM interfaces available:

- 1.5/2-Mbps T-1, E-1
- 6-Mbps J2 (Japan only)
- 25-Mbps UTP
- 34-Mbps UTP
- 34-Mbps E3 coax
- 45-Mbps DS-3 coax
- 51-Mbps UTP
- 100-Mbps TAXI fiber
- 155-Mbps fiber
- 155-Mbps UTP
- 622-Mbps OC-12 fiber

**TABLE 2.3 SONET
Specifications**

Data rate	Optical standard
51.84 Mbps	OC-1
155.52 Mbps	OC-3
622.08 Mbps	OC-12
1.244 Gbps	OC-24
2.488 Gbps	OC-48

Summary

Asynchronous transfer mode (ATM) is a switching technology that uses small, fixed-sized (53-byte) cells. It's a connection-oriented technology that is self-routing. ATM's bandwidth is scalable between 25 Mbps to 2.5 Gbps. The ATM Forum has defined several important ATM interfaces, including a user-to-network interface (UNI) between end users and ATM services and a network-to-network interface (NNI) between two switches or between two networks.

The ATM protocol reference model includes an ATM adaption layer responsible for segmentation and reassembly of data, an ATM layer responsible for routing information, and a physical layer responsible for transmitting and receiving data.

The ATM Forum has developed the LAN emulation user-to-network interface (LUNI) to fool LANs into thinking that they're communicating with another LAN when they're actually communicating with ATM equipment. Ethernet and Token Ring topologies are emulated, while FDDI LANs must be converted to Ethernet packets first. Devices that communicate with devices attached to LANs are known as LAN emulation clients (LECs), while LAN emulation servers (LESs) are responsible for resolving MAC and ATM address mappings using the LAN emulation address resolution protocol. A LAN emulation configuration server (LECS) is used to locate the address of the appropriate LES for an LEC seeking to establish a connection.

ATM network management specifications are still limited. A group of testing equipment vendors are working with the ATM Forum to develop specifications for monitoring ATM transmission. Security specifications for native ATM applications still need to be developed.

Permanent virtual connections (PVCs) must be set up in advance. The real breakthrough for ATM will come with switched virtual connections (SVCs) that are capable of being set up "on the fly" in real time.

Moving Toward Virtual Networks and ATM

Many of the legacy LANs discussed in chapter 1 are now experiencing serious traffic congestion problems as well as other problems related to their increasing complexity. In chapter 2, I focused on the technology associated with ATM and examined the concept of a virtual network. Chapter 3 brings these two worlds together and shows how real-life legacy LANs are beginning to incorporate ATM products.

Today Backbones Have Collapsed

The thick coaxial Ethernet backbone down the center aisle of a department has given way to unobtrusive 10BaseT, unshielded, twisted-pair wire. The backbone has "collapsed" into the hub. As Figure 3.1 illustrates, a hub's backplane can hold several circuit-card modules. This backplane is really the modern LAN's collapsed backbone. The collapsed backbone exists within the box that makes up the hub. Modules can add Ethernet, Token Ring, and FDDI topologies as well as bridging and routing functions to segment the LAN. The hub's backplane performs the backbone's traditional role of a high-speed communications channel linking together the various LAN segments within the hub. Many hub vendors are promising to incorporate ATM into their backplanes in the future.

Some Problems with Today's Legacy LANs

While the advent of 10BaseT and collapsed backbones have made it easier to manage network cabling, network managers of legacy LANs (networks with traditional frame-based topologies such as Ethernet, Token Ring, and FDDI) still face many problems. In chapter 1 you saw how new traffic-intensive data applications were filling bandwidth and decreasing network throughput. This trend is likely to continue

Ethernet LAN modules

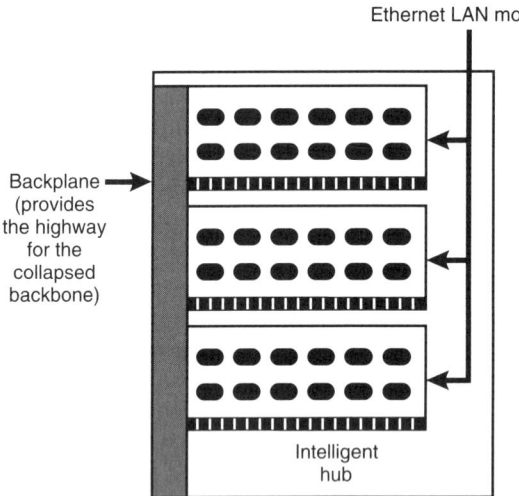

Backplane
(provides
the highway
for the
collapsed
backbone)

Intelligent
hub

Figure 3.1 A hub's collapsed backbone.

indefinitely and probably increase as more imaging and multimedia applications begin to appear on LANs. Some fast Ethernet vendors have positioned their products as the backbone bandwidth solution, but low sales seem to confirm that many companies are questioning the long-term adequacy of this technology for backbones. After all, with traffic increasing so dramatically, why settle for a dead-end 100-Mbps solution?

For a company with relatively small networks and Token Ring or Ethernet backbones, increasing performance can be as simple as adding a low-end switching hub. Figure 3.2 shows a typical corporate network environment. Here a LAN has been bridged into three segments. Unfortunately, the overall traffic on the backbone (It doesn't matter if it's collapsed or the traditional type) has increased to the point that throughput is no longer acceptable. Figure 3.3 shows the same network after the LAN segment with the most traffic has been split into three segments and attached to a switch. The file server originally on this segment has been given its own segment and switch port. Nodes on LANs 2 and 3 can still access the server on LAN 1, but without congesting the segments with the nodes attached.

Companies with a significant number of LAN nodes face a problem that's just as serious as increasing traffic and just as difficult to solve. Users are constantly moving from one desk to another. The "moves and changes" required of the network can tie up valuable network manager resources. Analysts believe that some key industries experience close to a 50% change in employee locations in the course of a year. As you'll see shortly, in situations such as this, it's much easier to keep track of users if they can be grouped together logically rather than physically. In other words, a virtual network makes sense both financially (less time devoted to moves and changes) and physically (trips to the wiring closet and additional time later cleaning up any inadvertent damage done by loosened cables).

The appeal of a virtual network has added attractiveness for network managers who have the TCP/IP protocol running on their networks. This protocol requires assigning an IP subnet as well as a node address to each node that joins the network or

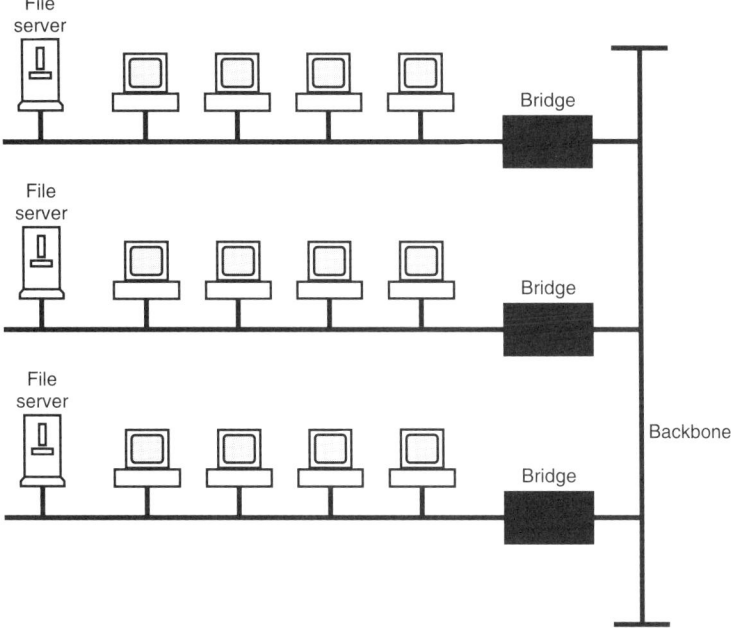

Figure 3.2 A backbone with too much traffic.

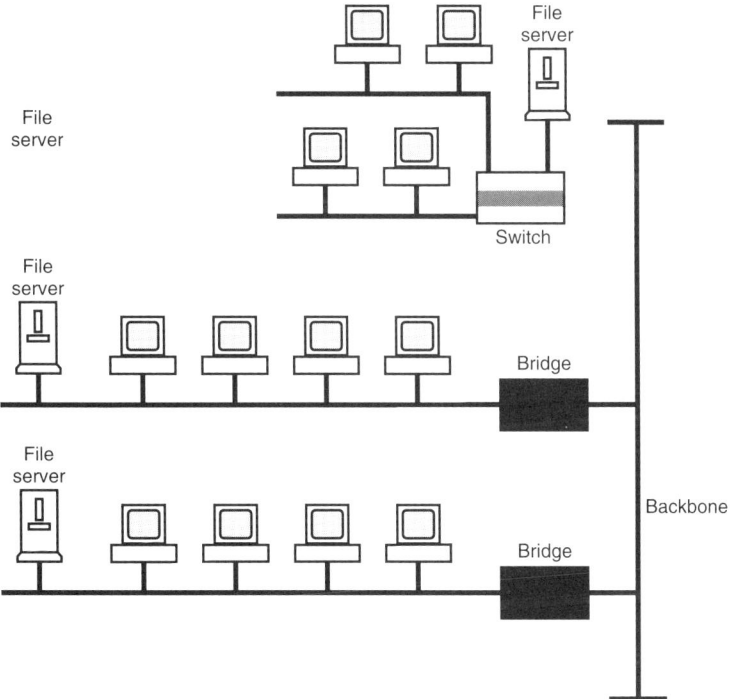

Figure 3.3 A switch introduced into the environment.

changes locations. As you'll see later in this chapter, some switches that create virtual LANs can automate the process of segmenting based on IP subnet as well as by node address.

Adding Switches to the Legacy LAN Environment

Many network managers have therefore begun to use Ethernet and Token Ring switches at the workgroup and departmental LAN levels to pull traffic off backbones and effectively segment LANs. The first-generation switching technology examined in chapter 1 is a form of layer-2 (of the OSI model) bridging. LAN switching can provide full 10-Mbps or 16-Mbps bandwidth (depending on whether it's Ethernet or Token Ring). It's lightening fast, but suffers from the same limitations as do all layer-2 devices. These limitations include the inability to route data based on their protocols and the inability to set up a "firewall" to screen out network broadcasts.

The Switching Hub and Virtual Networks

As vendors have incorporated switches into their intelligent hubs, a new product (the switching hub) has emerged. Suddenly it's possible to segment LAN users logically as well as physically for more efficient network performance. In chapter 2 you looked at ATM's role in virtual networks. The virtual networks you'll examine in this chapter are the precursors to true ATM networks. Some are designed only for legacy LAN environments, and others allow hybrid solutions that permit ATM switches to coexist with Ethernet switches. Many switch vendors now offer migration options to add ATM technology at a later date. To understand the strengths and weaknesses of specific vendors' ATM product plans (examined in part 2 of this book), it's absolutely crucial to spend some time looking at what virtual networking means to legacy LAN environments and the many different ways of achieving this goal.

Physical Versus Logical LANs

The bridges and routers we described in chapter 1 are physical devices that physically segment LANs. Historically, LAN managers have reduced network traffic by physically segmenting their LANs based on traffic patterns. Bridges might separate Accounting and Engineering LAN segments because users in these two areas generally share little of the same traffic. Under Ethernet's architecture, messages are broadcast to all users on a segment.

A major problem for network managers is that even when they segment users, they still devote a tremendous amount of time to moves and changes. A user in Accounting, for example, might move temporarily to another part of the building. Then network manager has to make the necessary physical wiring changes.

Figure 3.4 illustrates a traditional Ethernet LAN that has been segmented by bridges into two different LANs. Notice that you don't mix and match users in this situation. All the users on LAN 1 receive the LAN traffic routed to LAN 1. As shown in Figure 3.5, the physical LANs look very much the same. The difference is that a switching hub designates which users are grouped together to form a logical LAN

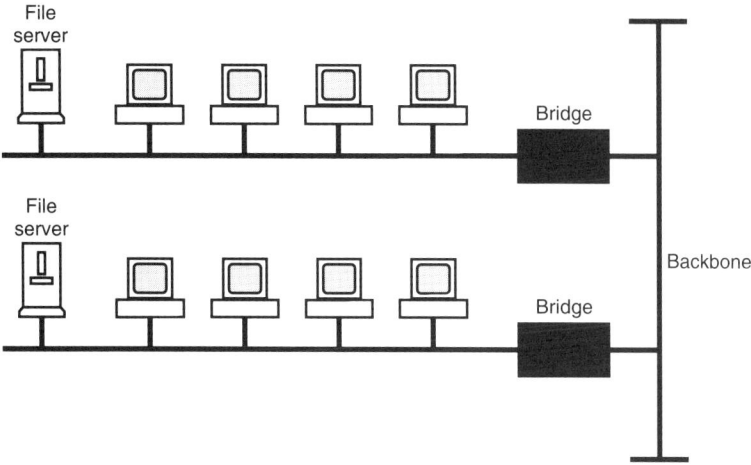

Figure 3.4 A conventional LAN.

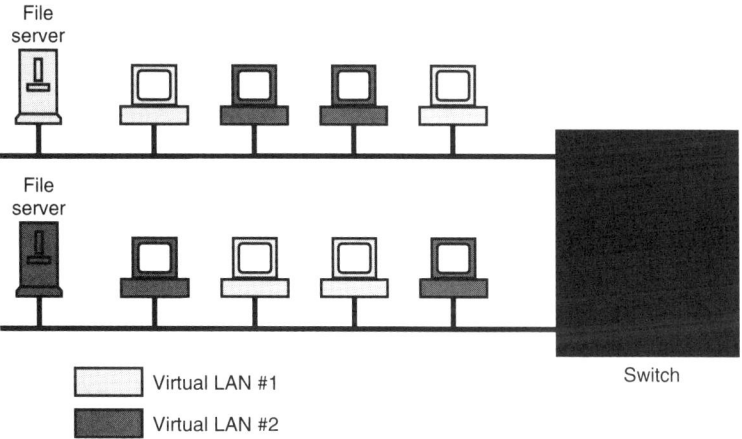

Virtual LAN #1

Virtual LAN #2

Switch

Figure 3.5 The same network as a virtual LAN.

segment. Another term I'll use for a logical LAN segment is a *virtual LAN*. There are at least three different approaches to creating virtual LANs. Perhaps the easiest method to understand, the *MAC address list*, consists of the network manager assigning node MAC addresses to a virtual LAN. (A number of hub vendors, including Cabletron, offer this approach.) The switch is given these MAC layer network addresses and transmits information to the appropriate user. The beauty of this logical or virtual LAN is that when users change assignments and need to work cooperatively on a project, they can be assigned to the same virtual LAN regardless of where they are physically in an organization.

Many of the Ethernet switches that create virtual LANs function at layer 2 (data link) of the OSI model described in chapter 1. Network managers use software to group together several ports into a virtual LAN. This LAN is given a unique number signifying its address. The Ethernet switch keeps a node's MAC address in memory, along with information on the port to which it's connected and the number of the virtual LAN to which it's connected. Each node on a virtual LAN receives all broadcasts directed to other nodes on the same virtual LAN. This type of virtual LAN is sometimes called a *virtual segment virtual LAN (virtual segment VLAN)*.

This approach creates a virtual LAN that's the equivalent to a single subnet that functions at layer 3 of the OSI model. What this means is that this type of virtual LAN creates only a single firewall for all protocols on that LAN and cannot differentiate between protocols. All LAN nodes with IP addresses in a virtual segment must have their subnet addresses changed to the same number. This type of virtual LAN is simple to administer because management consists of grouping together switch ports.

A far more complex type of virtual LAN is the *virtual subnet VLAN*. These types of virtual LANs are created by switches that can handle layer-3 subnet addresses for various protocols. The network manager assigns a subnet address to a number of switch ports that can be on different switches connected by a common backbone. Ports can be assigned a different subnet number of each protocol that's present. Nodes on a virtual subnet are treated as a bridge group, and traffic between virtual subnets is routed at layer 3.

Unfortunately, this approach doesn't work well for nonroutable protocols such as NETBIOS and LAT. The subnet ID found on each legacy LAN frame serves as its virtual LAN identifier within that virtual LAN. It serves as its network layer routing address for frames that must be routed from a different virtual LAN segment. This type of virtual LAN is not as fast as the virtual segment type because of the additional processing that must take place at layer 3. The addition of chips to facilitate MAC-level forwarding and a RISC processor for higher-level processing, however, help compensate for the requirements posed by its greater sophistication.

Agile Networks' Virtual Subnet VLANs

Agile Networks manufactures products that provide virtual subnets on virtual LANs. It defines a relational LAN as a group of geographically dispersed end stations having the same protocol type and subnetwork address. Such virtual networks are sensitive to both routable protocols such as IPX, IP, and AppleTalk and nonroutable protocols such as NetBIOS and DEC's LAT protocol. In a reiational network such as the one offered by Agile Networks, nodes running the same protocol in effect become part of the same broadcast group. In the case of Novell's NetWare (IPX protocol) this can be extremely useful because NetWare updates all servers using its service advertisement protocol (SAP) broadcasts. Because a relational switch knows which subnetworks contain only nodes used as clients, it can keep traffic off these subnets and broadcast to only subnets that need to have the information. Because nonroutable protocols can't be routed (what a surprise!), relational switches form one relational LAN for each nonroutable protocol.

For network managers looking for flexibility, the virtual subnet VLAN offers some very intriguing alternatives. In Figure 3.6, a LAN segment has end stations (or *nodes*, to use the equally common term) in three relational LANs, a NetWare net-

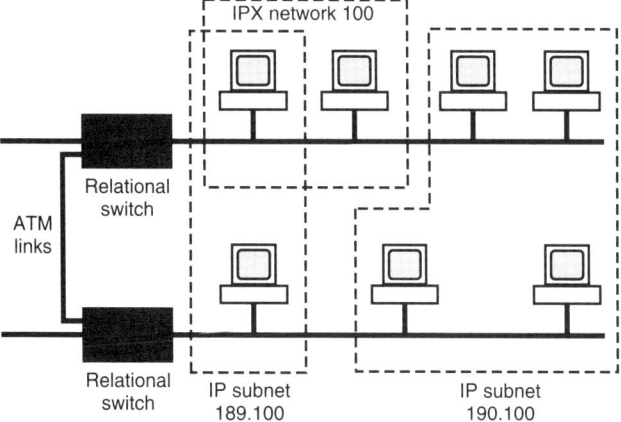

Figure 3.6 Agile Networks' relational switches create different subnets.

work (100) with IPX protocol, an Internet Protocol (IP) subnet 190.100 and a second network, an IP subnet 190.101. One node (marked with an asterisk) is a dual-protocol machine that belongs to two different relational LANs. Also note that segment 2 supports two different IP subnets.

IP Host-List Virtual LANs

Switches that permit the creation of MAC address lists create virtual LANs that are independent of switches and ports. The nodes can be moved form physical segment to physical segment and still be part of the appropriate virtual LAN. Unfortunately, they can't operate at the segment or subnet level and are not protocol-aware. A few vendors, including Newbridge Networks, have a variation of the virtual LAN I've already discussed. Lists of nodes' IP addresses are grouped together to identify virtual LANs. Nodes can be moved to any physical LAN segment, but they'll remain a part of their virtual LAN. All the nodes in an IP host-list virtual LAN must have the same subnet address to be switched; otherwise, it will be routed. This approach obviously has the advantage of being protocol-aware as well as allowing different virtual LANs to coexist on the same physical segment.

Rules-Based Virtual Networks

Xylan Corporation is the first vendor to indicate it has developed a switch to allow the creation of rules-based virtual networks. An AutoTracker feature within its switches allows network managers to define a virtual LAN based on a set of policies. The switches use user-defined frame fields as well as common subnet addresses, protocol type, etc. to create virtual LANs. A major advantage of this approach is that, since frame values can be used to represent whatever the network manager wants, they can be used to handle nonroutable protocols. It's even possible to define virtual LANs by packet type within an application so a single client could conceivably be a part of multiple virtual LANs.

Automated Virtual LANs

Perhaps the ultimate direction virtual LANs will take is toward greater automation to decrease the time requirements for network managers in administering the configuration. Today no vendor currently offers a product that's completely automated; all require varying degrees of a network manager's time. Agile Networks is moving in the direction of an automated product, but isn't there yet. Its ATMizer switch uses an agent technology that enables the switch to listen to all broadcast traffic in order to create a distributed subnet. Once virtual LANs are established and the mappings defined, the switch uses a proprietary protocol to broadcast this information over its backbone.

Incorporating ATM Backbones with Ethernet Switches

More and more vendors are beginning to offer hybrid Ethernet/ATM switches, switches that offer both Ethernet and ATM ports. Besides offering much higher bandwidth across a backbone for these switches, the ATM ports allow flexibility in backbone topology. This flexibility provides many advantages, especially additional fault tolerance. Figure 3.7 illustrates the problem with a traditional backbone topology for switches. A daisy-chain arrangement results in failure if one link is broken. Figure 3.8 illustrates a ring topology and a full mesh topology. Both topologies offer much greater fault tolerance. With a mesh topology, multiple failures won't bring down the network.

How Layer-2 Ethernet Switches Communicate Across a Backbone

Switches communicating across virtual LAN backbones is a confusing topic. Most people agree that, in the long run, sites with several switches and a number of virtual LANs will probably have ATM backbones, and each virtual LAN will then be able to have its own channel with which to communicate. At this time, though, vendors are using tra-

Figure 3.7   A backbone of daisy-chained switches can fail if one link fails.

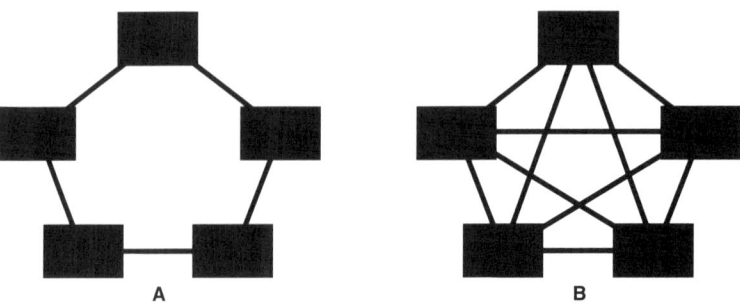

Figure 3.8   ATM switches with ring (A) and mesh (B) topologies.

ditional LAN shared media with detailed migration plans to ATM so customers don't feel they're stuck with dead-end technology. Today's virtual LANs are likely to have switches that communicate across a backbone using proprietary technology.

The process I described for creating and servicing virtual LANs works just fine with one Ethernet switch. There are some issues, though, when two or more such switches, each with their own virtual LANs, must be linked together over a shared-media backbone. One very primitive and expensive way to create backbones for multiple linked switches is to install parallel cabling. One set of backbone cabling would need to be installed for each virtual LAN that the network manager wanted to create. If three Ethernet switches were going to create four virtual LANs, then four sets of cabling would provide four different backbones, one for each of the virtual LANs. While some vendors advocate this approach, it's clearly not cost-effective and creates a morass of cabling that simply isn't manageable in a large network. There are several more sophisticated technological alternatives, however.

When switches need to exchange address information over a common backbone, one approach followed by a number of companies such as LANNET is to have a node transmit its address and virtual LAN information to its switch when it powers up. Its switch then sends a short message burst containing this data to other switches over the backbone they share. This approach creates overhead, particularly when several switches with lots of nodes share a backbone. Another issue is the need for the switches to synchronize their update messages.

A second method for switches to use to update each other is known as *tagging*. The Xylan company attaches a tag containing a node's virtual LAN identifier to its Ethernet frames. A major problem with this approach is that, if the data packet is already at Ethernet's maximum packet size, the tag cause the packet to break the protocol's rules concerning size. Companies like Xylan get around this problem by a using proprietary code, but customers must be careful to avoid mixing and matching switches from different vendors.

Still another approach for exchanging node address information between switches is to use time-division multiplexing to assign a time slot to a specific virtual LAN. Large networks can create the need for extensive bandwidth dedicated to this task, as well as a good deal of overhead.

Backbones for Layer-3 Switching Hubs with Virtual LANs

Imagine the same situation I described for Layer-2 Ethernet switches that create virtual LANs—several switches connected over a common backbone—with Layer-3 Ethernet switches. The techniques such as tagging and message transmission described for Layer-2 switching hubs won't work for Layer-3 Ethernet switches. Traditional routing algorithms also don't work in a switched virtual LAN environment. The solution used by such switches when they need to know how to route a frame is to broadcast or "flood" the virtual LAN and then record which port the responding frame comes in on. As such LANs grow larger, however, the increasing number of frame broadcasts can affect performance. In some cases, a process called *load balancing* is used to spread excessive traffic over different backbones. Special software from Novell permits NetWare servers to balance traffic over several different virtual LANs.

Routing in Today's Switching Hubs

Because most of today's Ethernet switching hubs function at layer 2 of the OSI model, they can't segment based on protocol nor can they route traffic from one virtual LAN segment to another virtual LAN segment. Traffic between virtual LANs requires a router, which can be a stand-alone multiprotocol box or router card module that's inserted within the switching hub. Unfortunately, nothing is free, particularly in the world of networking. As explained in chapter 1, routing is a much more complicated process than bridging, and as a result, when integrated with a switching hub, it slows down performance significantly. Just how much degradation it introduces is open to debate. While performance theoretically could decline by as much as 90 percent, vendors say that intelligent design of virtual networks to limit the amount of routing between segments to 20% or so, coupled with stripped down routing code, could slow performance by less than 10 percent.

ATM Edge Routers

NetEdge Systems has been in the forefront in introducing a new type of device, the ATM *edge router*, to facilitate communications between ATM devices and legacy LANs. The Agile ATM switch described earlier is an example of a device designed for handling traffic within virtual (relational) LANs while the job of routers, particularly the new edge routers, is to move traffic between virtual (relational) LANs. Routing is mandatory in an ATM environment because ATM switches don't provide end-to-end LAN connectivity.

Most of today's routers are not optimized to handle the high-bandwidth demands generated by an ATM backbone. NetEdge believes that the network environment that encompasses both ATM switched backbones and legacy LANs creates the need for small, modular routers that integrate ATM routing with LAN switching. Rather than traditional collapsed backbones that are linked to a large, centralized backbone router, NetEdge believes that these relatively inexpensive routers can be placed strategically to work in conjunction with a meshed network of ATM switches. This approach is superior, according to NetEdge, because it provides superior performance and also eliminates a single point of failure.

The edge router is particularly appropriate for companies with legacy LAN equipment that want to migrate gradually to ATM. NetEdge's ATM edge router creates virtual networks of nodes anywhere on the network, which means that a virtual LAN can be created with nodes on LANs with FDDI, Ethernet, Token Ring, and ATM. Servers can be placed on ATM networks for higher performance while clients are placed on lower-priced legacy LAN topologies. Figure 3.9 shows some ATM edge routers linking together two corporate locations via wide area ATM service. (I'll discuss that part of the puzzle in chapter 4 of this book.)

NetEdge's edge routers do a bit more than the LAN emulation examined in chapter 2. They map MAC addresses and subnet addresses to ATM addresses using a proprietary protocol. To locate a LAN IP subnet, the edge router sends out an address resolution protocol request on an ATM multicast channel dedicated to the internet protocol. An edge router that knows the correct address responds with this infor-

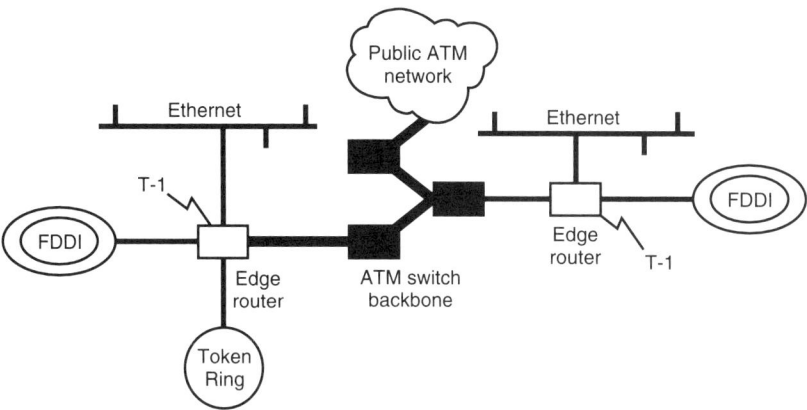

Figure 3.9 Edge routers in action.

mation. The edge router requesting the information caches the ATM address and then sets up a switched virtual circuit with the responding edge router using the Q.2931 signaling protocol across the user-network interface. Once that link is established, the remote router forwards the data requiring routing to the appropriate local subnet and node.

An Alternative to Edge Routers

In chapter 2, I described the LAN emulation specifications developed by the ATM Forum, explaining that they created layer-2 broadcast zones that are unaware of protocols and thus protocol differences. One issue not yet addressed is the lack of specifications for internetwork routing between emulated LANs; today that's the role of the router or edge router. The ATM Forum is working on a specification for ATM routing at the network layer of the OSI model. The specified architecture to support network-layer switching at all levels of an ATM-based network is known as *multi-protocol over ATM (MPOA)*. Eventually, MPOA will result in products that act as virtual distributed routers.

Under MPOA, route tables are distributed out to edge switches so they can make the appropriate routing decisions. A route server calculates route tables as well as routing rules and provides this information to edge switches. Figure 3.10 describes MPOA's architecture.

The route selection process is likely to be similar to the "next hop" approach of TCP/IP routing. The request will hop between route servers until a server is found that knows the ATM address. As in the case of LAN emulation discussed in Chapter 2, MPOA creates an ATM switched virtual circuit (SVC) when a data relationship like a TCP session is established. If the destination is not an ATM device, the ATM SVC's termination probably will be a router or hub connected to a legacy LAN which will format the data for the appropriate legacy protocol.

MPOA will permit direct communication between emulated LANs in much the same way that virtual LANs are handled by legacy LANs. It won't be easy to develop

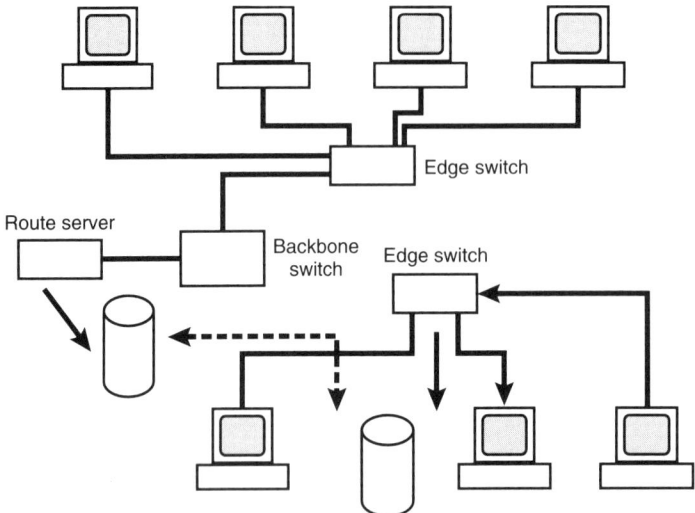

Figure 3.10 MPOA's architecture.

MPOA because it has to be able to support discovery protocols such as TCP/IP's address resolution protocol (ARP). Without this support, users would have to convert applications to run them on ATM. At this time some vendors, such as Net Edge Systems and Newbridge, have developed proprietary versions of such switches. Newbridge offers an ATM-to-Ethernet *ridge* (a hybrid bridge/router/switch). It also offers a route server that comes with the Sun Sparc20 platform.

Problems with Virtual Networks

As a step toward ATM, virtual networks sound ideal, but they do present several design problems for network managers. It's possible to have several LAN segments act as if they were on a single physical wire. The problem comes when there are nodes, likely those running IP protocol, on the same virtual LAN with different subnet numbers. Because layer-3 protocols assume that nodes with different subnet numbers are on different physical segments, frames from one node to a second node on the same virtual LAN but with different subnet numbers are likely to travel to a router and then back again. Because routing is a good measure slower than switching, it defeats the whole purpose of using a switch to create virtual LANs.

A similar problem results if a node is moved to a segment where the server has a different subnet address. Server and node will communicate, but they'll do so through a router even though they're on the same physical segment. To avoid this situation, network managers should try to ensure that every segment and node in a virtual LAN has the same subnet number.

Because NetWare's IPX protocol requires that each network interface card has its own subnet identification number, it automatically assigns a unique subnet number to each node on a LAN. This process doesn't work, though, when several segments comprise a subnet on a virtual LAN created by a switch. A NetWare server with several network interface cards will become confused enough to crash the LAN because it assumes that each of the NICs is on the same physical segment. Network managers should allocate one connection to a server per each virtual LAN. This connection often consists of a high-bandwidth pipe such as FDDI or 100-Mbps Ethernet.

What happens when more than one virtual LAN needs access to a server is that one virtual LAN will probably access the server via the high-bandwidth pipe while the unlucky second virtual LAN will be forced to go through a router. One possible solution is technology currently offered by 3Com and a few other vendors: NICs that allow multiple virtual MAC and subnet addresses on a single physical interface so servers can have different subnet addresses for each virtual LAN, even if it has only one high-bandwidth pipe.

Security is still another headache for network managers moving toward virtual networks. While security between virtual LANs can pretty well be handled by routers, what about security within a virtual LAN? Many vendors have developed proprietary network management software that permits filtering and blocking based on specific MAC addresses or segments.

In addition to security concerns, network managers interested in installing virtual LANs face possible serious interoperability issues with the *dynamic host configuration protocol (DHCP)* support in Windows 95. Layer-3 virtual LANs configured with nodes' IP addresses can potentially be at odds with DHCP, which automatically assigns IP addresses to nodes on startup.

When a virtual LAN is only IP-address-based, a node doesn't know what virtual LAN it belongs to or what DHCP server it should go to. While the goal of DHCP is to take configuration problems out of the hands of users and to relieve system administrators from the task of building tables that map MAC and IP addresses, this approach doesn't allow network managers to build IP address-based virtual LANs. It's possible to assign nodes to virtual LANs with layer 2; DHCP and virtual LANs are complementary under this scheme and DHCP can then use the MAC address to assign the node the appropriate IP address. Unfortunately, this approach this defeats the purpose of virtual LANs because it imposes physical limitations.

The Movement Toward VLAN Interoperability

Because vendors have developed proprietary methods for establishing and running their VLANs, there has been virtually no interoperability. Cisco Systems has been the prime backer of an IEEE 802.10 standard for VLANs that calls for routers to arbitrate interoperability across VLANs. Obviously, Cisco has a vested interest in a router-oriented solution. Other vendors, such as 3Com and Cabletron Systems, have urged a solution that would be centered in LAN equipment such as hubs and switches.

The IEEE 802.10 approach focuses on *frame tagging*, a feature that scans packet addresses before sending packets to their destinations. It would be particularly valuable for metropolitan area networks where security is crucial, by ensuring that packets reach the right destination. Cisco's competitors view 802.10 as a security and not a VLAN interoperability standard. The jury is still out on the final resolution of this dilemma.

Policy-Based Management of VLANs

Vendors vary as to the degree of policy-based management they plan to build into their VLAN products. Policy-based management enables managers to assign priorities to different types of application traffic. If an outage reduces bandwidth availability, for example, then the traffic with the highest priority will receive its necessary bandwidth. Within the near future, network managers should be able to develop policies that limit the time of day a user can be on the network and the amount of bandwidth allocated to that user. Similarly, network managers should be able to establish different topologies for different working conditions. A user might be part of one virtual LAN in the morning and a different virtual LAN in the afternoon because of a split job function. This type of flexibility could be automated.

Summary

Many physical backbones have collapsed into the backplanes of today's intelligent hubs. Bridging and routing takes place via add-on modules that perform these functions. First-generation hubs didn't solve congestion problems, though, because segmenting via bridges or routers does not increase overall bandwidth; it only reduces congestion on individual segments. Switches added to hubs save valuable time for "moves and changes." They make it possible to create virtual networks, networks that have a logical structure that might be far different than the physical structure of the network. This means that users can be grouped by their network needs and company functions rather than by where they sit. Virtual networks also increase aggregate bandwidth. These logical networks are particularly appealing for networks running the TCP/IP protocol suite because of the need to assign IP subnet addresses and note addresses that are subject to change whenever a user moves from one physical location to another physical location.

Layer-2 switches are lightning fast but can't route data or setup firewalls to screen out network broadcasts. You can create virtual LANs by using media access control (MAC) address lists to assign users to different virtual LAN segments. Virtual segment VLANs assign users to virtual LAN segments based on a switch's ports. Under this approach, different protocols cannot be differentiated from each other and assigned to different virtual LAN segments. The virtual subnet VLAN operates at layer 3 and can handle subnet addresses. It differentiates protocols and group users based on the protocols they're running.

Switch vendors have taken different approaches to creating virtual LANs. IP host-list VLANs are based on IP addresses, while rules-based virtual networks allow a network manager to define a set of policies that the switch follows in assigning LAN segments to nodes. Ultimately, an automated approach will use intelligent agents to permit the switch to determine LAN segment assignments.

More and more Ethernet switches will be linked via ATM backbones. The ATM switches are likely to be grouped by ring and mesh topologies rather than by a daisy-chain arrangement to provide greater system fault tolerance. Edge routers are devices designed to move traffic between virtual LANs. They're optimized to handle the high-bandwidth demands generated by an ATM backbone.

An alternative to edge routers could be an architecture to support network-layer switching at all levels of an ATM-based network. This architecture is known as multi-protocol over ATM (MPOA), and sends route tables out to edge switches so they can make routing decisions.

Today's network managers face several problems relating to using virtual LANs. Nodes running the IP protocol on the same virtual LAN segment with different sub- ɔ have their traffic routed, slowing down performance. If a th a server and they have different subnet addresses, the routed. Another major concern to network managers is the LANs with Microsoft's dynamic host configuration protocol rted in Windows 95. The current "work-arounds" that are le purpose of creating a virtual LAN.

Size _____

Style _____

Qty. _____

Dkt No. _____

Bundle No. _____

Wide Area Network Basics and ATM

Even though ATM was originally designed as a wide area network technology, it has made much more progress at the LAN level. The ATM Forum must still grapple with several ATM wide area network interface issues. In order to understand the role of ATM in wide area networks, it's necessary first to understand how other wide area network technologies work. Many customers who have already leased T-1 and T-3 lines or frame relay services are starting to wonder about migration paths to ATM. This chapter is as a primer for wide area networks. You'll learn about T-1 and T-3 lines as well as the principles behind frame relay service. You'll then see how the ATM Forum is moving toward interfaces that will make it possible to add ATM to the mix of current WAN technologies so customers can have the best of both worlds.

What Is a Wide Area Network (WAN)?

A *wide area network (WAN)* is a network that links together networks located in different geographical areas. A company might have to link together its Ethernet LAN in Boston and its Token Ring network in Los Angeles. Another company might want to link together several sites so they can exchange voice, data, and even video information.

One major problem for network designers has been that, while PC-based local area networks and mainframe networks both routinely transmit data at several million bits per second, transmission speeds over phone lines are considerably slower. Linking two 10-Mbps Ethernet LANs with a 19.2-Kbps analog phone line is bound to create a serious traffic bottleneck. Many companies still struggle with 9,600-bps modems. This congestion problem is becoming more manageable because of the growth of digital transmission services, such as integrated digital services network (ISDN), and also because of the development of more sophisticated types of services such as frame relay. I'll describe frame relay a bit later in this chapter.

Unreliability of WAN Links

A major problem for network managers and network designers is that WAN links are inherently less reliable than local transmission links. Error checking might not be crucial on a Token Ring LAN, but it becomes a major concern when data is transmitted over phone lines that are subject to all kinds of electrical interference. When remote communications are disrupted and data has to be retransmitted, the LAN-to-WAN link becomes even more congested and inefficient.

Divestiture and Wide Area Networks

The breakup of the Bell system dramatically changed the telecommunications industry and, in turn, how companies transmit their voice and data on wide area networks. Prior to divestiture, companies dealt exclusively with Bell. Today, network managers as well as systems integrators designing wide area networks must deal with several different vendors. From the customer's premises to the closest central office (CO) is the province of the local phone company, a *local exchange company (LEC)*. LECs include the Bell operating companies (Pacific Bell, Bell Atlantic, Southwest Bell, etc.) as well as a number of independent companies.

Here's where the telecommunications jargon becomes a bit overwhelming, but it's important to understand who you must deal with in order to send voice or data from one network to another network. The LECs control all calls made within their geographical areas or, *local area transport area (LATA)*. An LEC provides service within all the LATAs within its territory, but it cannot route a call from one to another without going through an interexchange carrier's network. Within each LATA are interface points to the inter-LATA carriers, known as *point of presences (POPs)*. Each inter-LATA carrier such as AT&T, MCI, Sprint, etc. have their own lists of POPs. AT&T calls its POP a central office (CO). These POPs are the only places within a LATA where an inter-LATA carrier can receive and deliver traffic. Figure 4.1 illustrates the route a call takes from one LATA to another.

Digital Signal Transmission

In 1957 the Bell system installed its first T-1 trunk to carry high-speed digital voice signals. Since data communications means digital and not analog signals, a device is needed to generate these digital signals. The digital pulses generated at a customer's site have to be filtered effectively to eliminate noise and distortion. Customers interface with the telephone company's digital network through a *channel service unit (CSU)* or *a data service unit (DSU)*.

Both the CSU and the DSU function as digital modems, though the CSU also provides some line conditioning and diagnostic functions. A *T-1* trunk contains 24 channels, and each channel is capable of handling 64,000 bits per second of transmission. An additional 64,000 bps is required for error checking, so one T-1 line requires a bandwidth of 1.5444 Mbps:

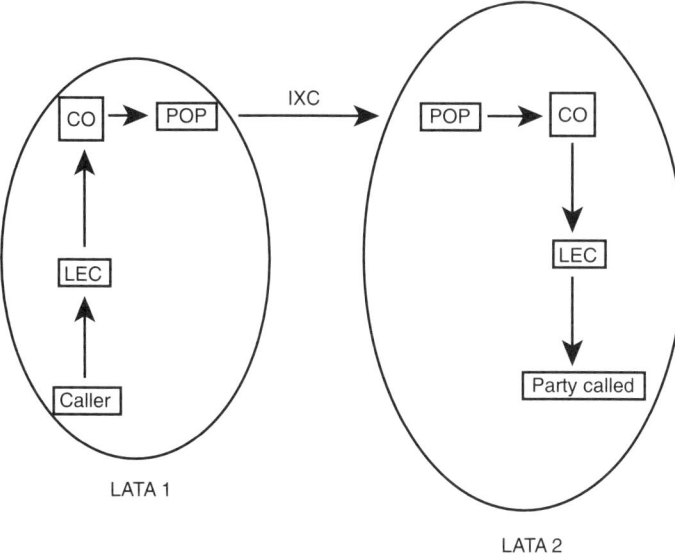

Figure 4.1 The route a call takes.

$$64,000 \text{ bps per channel} \times 24 \text{ channels} = 1.536 \text{ Mbps}$$

$$8 \text{ bps per sample} \times 8,000 \text{ samples per second} = 64,000 \text{ bps}$$

$$\text{Total} = 1.544 \text{ Mbps}$$

This 1.544-Mbps rate is known in the telecommunications industry as *DS-1 (digital signal level 1)*. There's an entire hierarchy of digital signal bandwidth options available, as shown in Table 4.1.

The T-1 (DS-1) and T-3 (DS-3) circuits are popular. The European digital hierarchy is slightly different from this North American standard. In Europe, DS-1 (E-1) consists of 32 channels, each of which transmits at 64 Kbps for a total bandwidth of 2.048 Mbps. Two of these channels are used only for signaling and network management functions.

TABLE 4.1 The Digital Signal Bandwidth Hierarchy

Signal	Speed	Number of T-1 channels
DS-1	1.544 Mbps	1
DS-2	6.312 Mbps	4
DS-3	45 Mbps	28
DS-4	274 Mbps	168

Why T-1 Circuits Are Popular

There are a number of reasons why T-1 (and by extension T-3) circuits are so popular. Digital transmission produces much higher-quality voice signals than analog transmission. As mentioned earlier, companies can save substantially by consolidating their voice and data transmissions over the same circuit rather than by maintaining two separate transmission paths.

Many T-1 multiplexers provide redundancy by offering automatic alternate routing so other circuits can be used if a path is out of service. Another advantage of T-1 service is its flexibility. The bandwidth can be allocated in different ways depending on voice and data needs. For example, a typical multiplexer might permit users to program data channels from 300 bps to 1.5 Mbps. You can also program voice channels to transmit at 64 Kbps or at 24 or 32 Kbps by using different compression schemes.

T-1 multiplexers usually have both synchronous and asynchronous data interfaces. A network manager with several low-speed data transmission lines could take advantage of a multiplexer to consolidate them into a single DS-0 channel.

T-1 frames

A T-1 trunk carries a serial bit stream that's transmitted using time-division multiplexing on a frame-by-frame basis. A frame consists of 192 bits (8 bits × 24 channels) plus one additional synchronization bit, for a total of 193 bits.

Time-division multiplexing: the key to T-1 transmission

Data from a number of different sources feeds into each of the 24 DS-0 channels composing a DS-1 line. This data must be placed in the appropriate frame format and then transmitted via a T-1 multiplexer using *time-division multiplexing (TDM)*, which consolidates the data stream from each of the DS-0 channels by using an approach that guarantees a time slot for data from each of these channels. As Figure 4.2 illustrates, it's possible for a channel to not have anything to send when its turn arrives; under such circumstances the time slot remains empty.

Channels

1		Asynchronous data			
2	Synchronous data	Synchronous data			
3		Fax		Fax	
4	Voice	Voice			
5	Voice	Voice	Voice	Voice	
6	Asynchronous data		Asynchronous data		
7	Synchronous data	Synchronous data			
8	LAN data	LAN data		LAN data	

Figure 4.2 Time-division multiplexing.

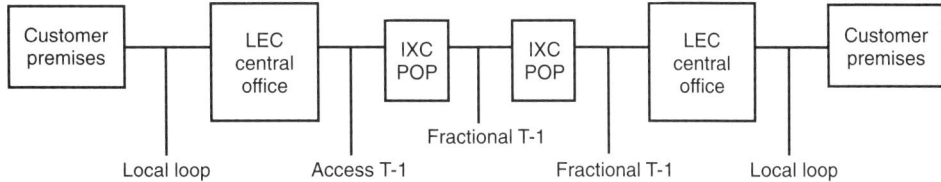

Figure 4.3 Using fractional T-1 lines.

Fractional T-1

Common carriers can subdivide the T-1 bandwidth into its 24 DS0 channels of 64 Kbps. Users can then bundle together these channels to fit particular applications. A customer might require a *fractional T-1* consisting of one eighth (192 Kbps), one fourth (384 Kbps), or one half (768 Kbps) of the available bandwidth. Users pay for only the bandwidth they need.

It's possible to compress voice transmission a good deal without losing any content (think of all the "ahhs" and "uhhs" in a typical conversation). With popular compression schemes, it's common for customers to take a 384-Kbps FT1 channel and use it for 20 voice lines of 16 Kbps each, or perhaps 10 voice lines of 32 Kbps apiece. Because of the efficiency of these compression schemes, no meaningful content is lost. Figure 4.3 shows a typical use of fractional T-1 lines.

T-3 multiplexers

Some companies use T-3 multiplexers to consolidate T-1 and fractional T-1 networks. It's definitely economical to use T-3 circuits rather than multiple T-1 circuits. If multiple T-1 lines are going to the same destination within a pathway, they can be combined and transmitted over a T-3 circuit. For relatively short distances, many industry experts use a 5-to-6 cost ratio for very short distances and an 8-to-10 cost ratio for longer distances. This means that it's cost-effective to use a T-3 line to replace 5-to-6 T-1 circuits. The added bandwidth is virtually free and can be used to carry additional voice traffic. Some companies that have traditionally used separate lines for voice and data have consolidated these two different streams using a T-3 line.

Packet-Switched (X.25) Networks

One of the major advantages of international standards in data communications protocols is illustrated by public data networks. These networks switch data in the form of CCITT X.25 packets across the country and around the world. The CCITT X.25 recommendation is based on the first three layers of the OSI model.

A *packet assembler/disassembler (PAD)* provides the protocol translation from a data stream's native protocol (such as an IBM mainframe's SNA protocol) to the X.25 protocol. At the destination end of the transmission, the PAD translates the packets from X.25 protocol back to whatever protocol is required. PADs help make X.25 transmission economical by concentrating several different data streams.

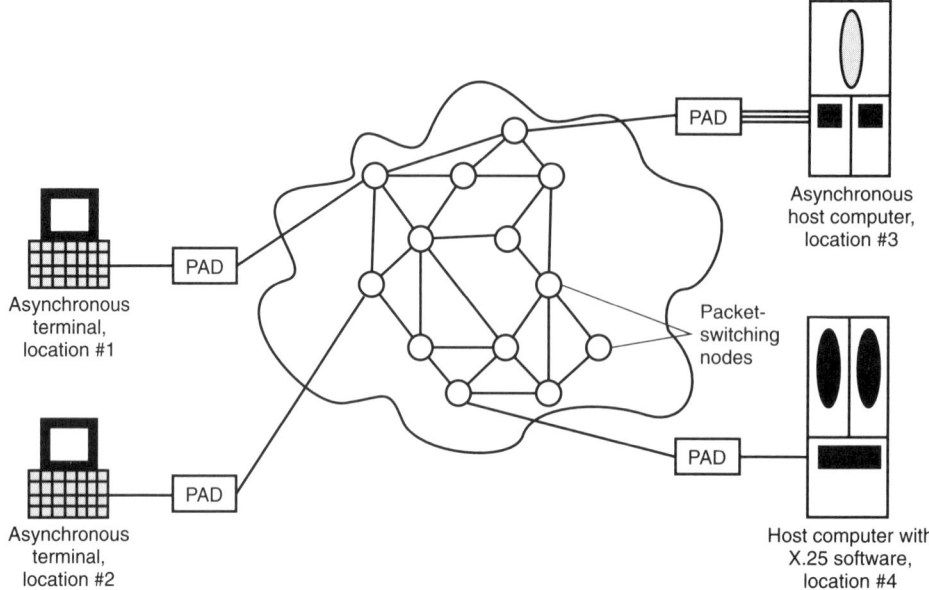

Figure 4.4 A packet-switched network.

The PAD must place the data it receives in a packet that contains control information for error checking and sequencing. Packets have a destination address; at each stage of their transmission route, they're checked for errors before being forwarded using the best available route at that particular moment. If a node receiving a packet detects an error, it requests that the sending node retransmit it. The path an individual packet takes is determined by the switching equipment, and no two packets will take the same route. As Figure 4.4 illustrates, it's not just likely but probable that packets will arrive at their destination out of sequence and need to be placed back in the proper order. All this activity is transparent to users. Note that packet-switched networks are usually depicted in diagrams as clouds.

One key consideration in considering packet-switched networks is that the cost of using such a network is based on the amount of information transmitted and not the distance between locations. Cost is usually based on the number of packets sent.

X.25 networks are also appealing for companies that have geographically dispersed heterogeneous networks. A company might have an SNA network at one location, for example, and a pre-SNA network at another site composed of pre-SNA bisynchronous equipment. A third site might have Digital Equipment Corporation (DEC) equipment running a VMS operating system. The protocol conversation required for communications among these disparate networks can be handled by an X.25 network vendor.

Private packet-switched networks

Some companies might need a packet switched network in an area where there are no public packet-switched services. It's also possible that they might have enough traffic between two sites to generate substantial tariff charges, and would benefit by building their own private network. A third advantage to a private packet-switched network is that it provides centralized management and control, as well as increased security since the company has total control over all traffic on the network.

X.25 Packet switching versus T-1 multiplexing

It can be a bit confusing differentiating between the two major methods of transmitting data on a wide area network. You've already looked at the T-1 multiplexer and the basic technique it uses, time-division multiplexing. Each channel on a time-division multiplexer is allocated a portion of the bandwidth, which is totally dedicated to that particular channel. T-1 lines are cost-effective only up to a certain distance. Since a company pays for the use of a line 24 hours a day, the higher the usage, the more cost-effective the transmission link becomes.

Packet-switching, on the other hand, uses a technique known as *statistical multiplexing*, which means that bandwidth isn't permanently allocated to any given channel; it's dynamically allocated with statistical algorithms, providing bandwidth to each channel based on that channel's need at any given time.

This ability to dynamically allocate bandwidth is particularly efficient when dealing with "bursty" traffic. A fax transmission, for example, might require massive amounts of bandwidth, but then that channel might remain idle for a substantial period of time.

Every advantage has its price, however. In order to be able to dynamically allocate bandwidth, a packet-switching statistical multiplexer must maintain constant communication with all channels to monitor what the channels needs are; this communication requires overhead and causes some delay time.

If data flow is reasonably constant on different channels, then time-division multiplexing has the edge. T-1 multiplexers were designed to handle voice traffic, and it should come as no surprise that they excel in this area. The various compression schemes used with voice traffic still require a 64-Kbps channel; they simply provide more voice conversations over this channel.

A T-1 multiplexer requires less overhead and can be considerably faster. T-1 multiplexers aren't sensitive to protocols because they simply plug their control information into a preassigned slot and don't worry about the protocol that the data uses.

What packet-switched networks do well is dynamically allocate their resources based on need and then route the packets efficiently. If a virtual circuit is tied up, they simply route packets along a different circuit. Since public packet-switched networks generally charge by the packet, packet switching can be much more cost-effective for transmitting small amounts of data. Also, many services offered by packet-switched networking companies, including protocol conversion and encryption, can help a company overcome incompatibilities between sites while maintaining security.

Creating a WAN by bridging LANs via X.25

Imagine a company with several sites scattered across the country. These sites have Ethernet LANs, but the network operating systems vary and include an IBM LAN server, Novell's NetWare, Apple's AppleShare, and Banyan VINES. The sites are scattered throughput the country, so the cost of T-1 links would be prohibitive.

As Figure 4.5 illustrates, you could link the sites together by using X.25 Ethernet bridges; since the LANs all use Ethernet, using X.25 remote bridges eliminates any concern for the different higher-level protocols running on these networks.

Let's consider a different scenario for a WAN. Instead of all the LANs having the same media access, assume that there's a variety of different access methods but the same upper-level protocol, in this case NetWare and IPX. Say the company has its corporate headquarters in Dallas; regional branches at 35 sites, including New Orleans, San Francisco, and Chicago; and Canadian headquarters in Montreal. Each site has its own NetWare LAN. These LANs use whatever media access method is cost-effective at that particular site, including Ethernet and Token Ring. The company needs to link all sites together for electronic mail, LAN maintenance, and updating key information, including scheduling. A major problem for this company is that many of the sites don't have local public packet-switches services available.

As Figure 4.6 illustrates, this particular company can combine a number of different technologies to build its WAN. Each site's NetWare LAN uses a communications file server with an X.25 gateway card that can transmit up to 64 Kbps. Each site uses

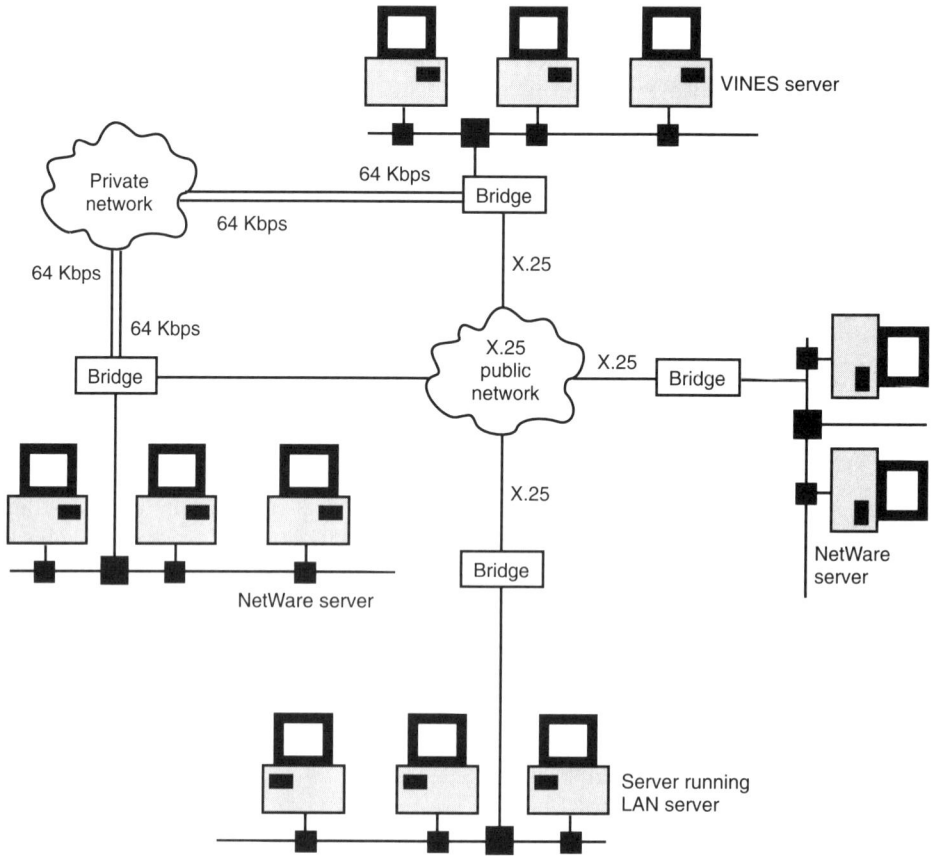

Figure 4.5 A WAN with X.25 Ethernet bridges.

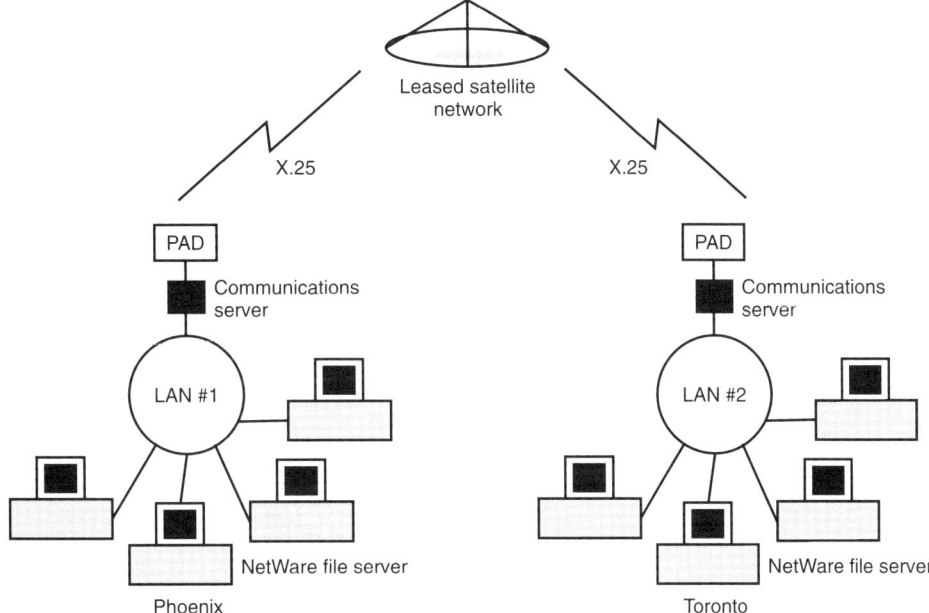

Figure 4.6 A WAN incorporating LANs and satellite transmissions.

a multiplexer to transmit the packets to a modem, which in turn transmits the signals to an RF (radio frequency) format that's transmitted to a satellite dish, which
uplinks the information to a leased satellite. The satellite serves as a backbone connecting all sites.

Fast-packet technology and frame relay

Fast-packet technology is a relatively new development in T-1 transmission that
combines the best features of T-1 multiplexing with some of the advantages of
packet switching. Special T-1 mutiplexers generate fast packets designed for a single channel only. The multiplexer allocates bandwidth instantaneously based on the
data streams it receives. Let's assume that a multiplexer must handle voice, LAN
traffic, and fax transmissions. This type of data mix would really benefit from a multiplexer that could allocate a single very large channel for a short period of time to
expedite the transfer of a large amount of urgent data.

Besides the ability to dynamically allocate resources based on need, another reason why fast-packet technology transmits packets that are truly "fast" is that, when
dealing with voice transmissions, this type of multiplexer frames voice bits only and
filters out the silence. It then compresses what's left and repackages the bits into
packets. Each packet contains an address up front. Notice that fixed time slots and
fixed channel allotments aren't required because the fast packets have addresses indicating where they're destined.

Another major advantage of fast-packet technology is that the multiplexer can be
programmed to recognize which data streams are crucial and send them first. Voice

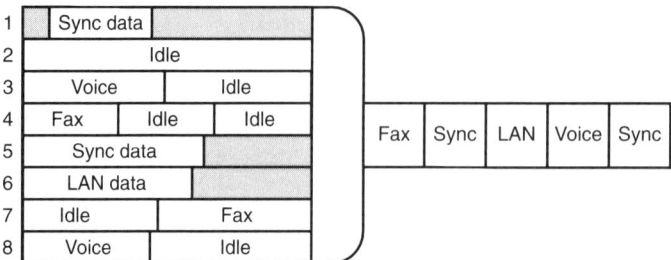

Figure 4.7 Fast-packet technology.

information, for example, might be given higher priority than a terminal's transmission. Figure 4.7 illustrates fast-packet technology in action.

While X.25 packet-switched networks use layers 1, 2, and 3 of the OSI model, fast-packet technology uses only the first layer and a portion of the second layer. Since the frames contain their own control information for source and destination addressing and error detection, the fast-packet switches need to look at only the destination addresses and then pass the frames along quickly. Rather than include requests for retransmission of data with errors, the destination node simply discards data that contains errors, which results in less processing and therefore increased speed. While X.25 has a maximum transmission speed of 64 Kbps, fast-packet technology can provide a rate of up to 2 Mbps.

Frame relay technology

A new standard has recently evolved for using fast-packet technology more efficiently. *Frame relay* is a data link layer protocol that defines how variable-length data frames can be assembled. Frame relay requires only 48 bits of overhead, four to five times fewer bits than required with packet switching. Frames can be sized appropriately for the data loads they need to carry. The frame relay standard provides for a *data link connection identifier (DLCI)*, which serves as an address field to allow for multiple logical sessions per physical data link. A standard frame format is specified for each of the various packet subsystems that will receive packets. Frames are labeled with the appropriate DLCI so, when the fast-packet system receives them, they can be sent to the appropriate destination. Error detection is performed only at the destination.

Fast-packet technology using the frame relay approach can handle massive amounts of packets efficiently because it allocates bandwidth dynamically. Northern Telecom has added frame relay to its switches, which enables carriers to offer fame relay as a service. Frame relay offers greater efficiency than X.25 technology because it combines the statistical multiplexing and port-sharing functions found in X.25 packet switching with the high speed and low delay of time-division multiplexing.

As frame relay technology matures, new specifications from the frame relay forum as well as add-on services from carriers are making this technology more attractive for network managers who are developing enterprise networks that incorporate WANs. The frame relay forum has developed a switched virtual circuit (SVC) standard. Until January 1994, frame relay communications services were limited to *per-*

manent virtual circuits (PVCs), which permitted communication only within certain geographic areas or business entities. With PVC, when ordering a circuit, users must tell the carriers the point of origin and the virtual path's destination. Then the carriers must make any changes to the PVCs. The SVC standard expanded frame relay has affected wide area networks because it permits dynamic switching of frame relay connections across geographical regions in a seamless manner. SVC users have much greater control of communications.

Switched Multimegabit Data Service

To confuse network managers even more, carriers offer *switched multimegabit data service (SMDS)*, which competes directly with frame relay. SMDS, unlike frame relay, offers connectionless data services that require no call setup or teardown procedures. All attached nodes can be connected on the fly. SMDS's connectionless nature permits a virtual LAN to be created over a metropolitan area. SMDS service is available at several different speeds, ranging from 1.14 to 45 Mbps, but this service isn't available in all areas.

SMDS offers considerable flexibility in connecting many different sites where dedicated links would be prohibitively expensive. The service is ideal for smaller organizations with high-volume data needs that can't afford dedicated private networks. SMDS is designed for applications that consist of frequent but short transmissions, such as electronic mail or limited database access. Network managers can screen addresses to create their own virtual networks in which only authorized participants can receive or send data, and it's also possible to broadcast data to multiple recipients.

SMDS is based on the IEEE 802.6 protocol for transmission over metropolitan area networks. The 802.6 packets are cells with a fixed size of 53 bytes, five of which are header information. Because of the similarity of its cell structure with that of ATM cells, it's likely that SMDS networks will eventually become components of wide area network ATMs as they develop.

Integrated Services Digital Network (ISDN)

Until recently, *integrated services digital network* was still jokingly referred to as "it still does nothing (ISDN)." Today there finally appears to be some real progress. ISDN is an evolving set of international standards for connecting voice, data, and video equipment. An ISDN user can carry on a voice phone call while also viewing video images or retrieving data from a computer. All these different types of information can travel in a single ISDN interface circuit packet. ISDN interfaces between local exchanges and end users could replace some T-1 trunks.

Basic rate interface (BRI)

The *basic rate interface (BRI)* refers to a single access point into ISDN. Known as 2B+D, BRI consists of two bearer channels and one data channel. Each bearer channel operates at 64 Kbps and is a clear channel, meaning there's no restriction on the format or type of information that passes through it. The data channel operates at 16 Kbps and is used for signaling and control information.

Primary rate interface (PRI)

The *primary rate interface (PRI)* allows multiple users to connect to ISDN. It will be used primarily to connect PBX, LAN, or another multiuser switching device to an ISDN network. The North American standard, followed by the United States, Canada, Mexico, Japan, and South Korea, consists of 23 B channels of 64 Kbps each and one D channel of 64 Kbps. The aggregate capacity is 1.5444 Mbps, or the equivalent bandwidth of a T-1 facility. T-1 is intended to be the chief facility used with the North American standard for PRI. The European standard for PRI consists of 30 B channels and one D channel, for an aggregate capacity of 2.048 Mbps. Because of the greater capacity available under PRI, it supports an additional channel, H. Three types of H channels are specified:

- H0 (384 Kbps)
- H11 (1.536 Mbps)
- H12 (1.920 Mbps)

The North American standard incorporates H0 and H11 channels, while the European standard incorporates H0 and H12 channels.

ISDN equipment

There are several different types of ISDN equipment. TE-1 has ISDN-compatible terminal equipment, which can be connected directly to an ISDN network. Unfortunately, many companies already have substantial investments in non-ISDN equipment. These products are referred to as TE-2 equipment because they require an interface device known as a *terminal adapter (TA)*. A TA can convert signals from an international standard, such as RS-232C, to the ISDN standard. There are other types of ISDN equipment designed for the local exchange companies and common carrier's networks, topics I won't discuss in this book.

Integrated voice under ISDN: IEEE 802.9

The IEEE 802.9 Committee has been working for several years to develop a set of specifications for a standard to support voice and data transmitted over a single network. The goal is to ensure that this new standard will provide IEEE 802 MAC services while being ISDN-compatible.

The *IEEE 802.9* standard will cover communication between an integrated voice/data terminal (IVDT) and a LAN, where the IVDT's access unit (AU) provides all required services. It will also cover a situation in which the services required for an IVDT to serve as a gateway to a backbone network.

The 802.9 standard must be able to provide multiple access to a single wire. It does so by using time-division multiplexing and assigning a block of time to each channel. The IEEE 802.9 standard will include a P-channel protocol at the MAC level designed to handle packet mode or burst data. The link access procedures on the D channel (LAPD) protocol used by ISDN at the MAC level is also used by 802.9 networks. This protocol ensures that 802.9 integrated voice/data networks will be able to use ISDN D-channel control information.

The 802.9 specifications are now a standard. The success of 802.9 depends on the success of ISDN. If ISDN is successful, 802.9 will provide the "missing link" for a LAN's MAC layer protocols and the protocols associated with ISDN. It will be the glue that makes simultaneous voice and data transmission possible between a LAN and an ISDN integrated voice data terminal.

ISDN in action

The common carriers have already targeted key corporations for ISDN. McDonald's serves as an example of how customers can use ISDN. The company has developed plans with AT&T and Illinois Bell for a global ISDN network. The company's plans include a combination of 2,100 BRI and PRI lines at its corporate headquarters to link it with 40 domestic field offices, more than 8,000 U.S. stores, and 3,000 overseas stores. A major advantage of ISDN for McDonald's is that it will be able to consolidate 21 existing telecommunications networks into one. A user at any ISDN terminal will be able to access database information anywhere on the network.

Their use of ISDN includes services available on an ISDN network such as electronic directory, which includes caller identification, message waiting, and message retrieval. Caller identification displays the number of the person calling from within the company and that person's name. Message waiting alerts users when a message has been left for them, and message retrieval enables users to examine a series of waiting messages by scrolling through them.

How long until an ISDN world?

While ISDN has made enormous progress the past few years, it isn't quite all here yet. The problem is that the regional Bell operating companies (RBOCs) have been deploying ISDN at very different rates. While it's very easy to get ISDN service in Northern California, for example, it could be very difficult in a small town in Texas. If a company has branch offices that span several different states, it might find that it has problems creating a wide area network that's all ISDN because of the lack of uniform deployment.

Synchronous Optical Network (SONET)

In 1986, the *SONET (synchronous optical network)* standard was introduced to synchronize public communications networks and then tie them together via a high-speed fiber-optic links. SONET defines a set of framing standards that determine how bytes are transmitted across links. What SONET offers is enormous bandwidth based on multiples of the base rate (OC-1) of a 51.84-Mbps or one T-3 link. The U.S. originally proposed a 51.84-Mbps signal, while Europeans wanted a 34-Mbps signal. The two groups compromised at a basic rate of 155.52 Mbps, or OC-3, to connect the two groups' signals. Table 4.2 provides the SONET hierarchy of transmission options.

Since SONET is built on the foundation of T-3, it's crucial that companies thinking of purchasing T-3 multiplexers in the immediate future receive a commitment from their vendors that the multiplexer is migratable to the emerging SONET standard.

TABLE 4.2 The SONET Hierarchy

Optical carrier number	Transmission rate (bps)	T-1s	T-3s
OC-1	51.84	28	1
OC-3	155.52	84	3
OC-9	466.56	252	9
OC-12	622.08	336	12
OC-18	933.12	504	18
OC-24	1244.16	672	24
OC-36	1866.24	1008	36
OC-48	2488.32	1344	48

The device must have sufficient backplane bandwidth to support at least SONET's OC-1 and ideally enough bandwidth to support multiples of OC-3 for future growth. While the SONET equipment that's already available will be used mostly for public data networks, some companies, such as Apple, are already using SONET for their own private networks.

SONET's phase II will provide operational and administrative support, including maintenance information of the data traveling over the links. Phase II will define the full seven-layer OSI protocol stack, including protocols associated with flow control to prevent data collisions. Among the carriers offering SONET are AT&T, MCI, and the regional Bell operating companies.

ATM and T-1 Links: AIM

There are likely to be many situations in the near future when network managers will want the bandwidth possible with T-3 lines (45 Mbps) but will be unwilling to pay ten times the cost of a T-1 line for 28 times the bandwidth. The solution might lie in connecting two locations using multiple point-to-point T-1 links and manage them as a bundle so they can avail themselves of ATM services and technology. The ATM Forum is working on an approach to achieving this goal, known as *asynchronous transfer mode inverse multiplexing (AIM)*. Under AIM, it would be possible for companies to connect ATM equipment and services to as little as a T-1 ATM link, providing it has an AIM interface. The AIM specification is likely to be approved by the ATM Forum sometime in 1996.

How AIM will work

Figure 4.8 shows how AIM will operate. Under AIM, a switch's ATM layer would pass cells to the AIM's *transmission convergence (TC)* sublayer in either the user-to-network interface (UNI) or the network-to-network interface (NNI) formats. The TC sublayer uses the AIM protocol known as ACP to arrange cells in the delay compensation buffer so the receiving end can recreate the original cell stream. The TC sublayer can then transmit the cells to the multiple T-1 links using a round-robbin approach. The cells travel over the various T-1 links until they reach their destination ATM switch. The TC sublayer receives the cells and compensates for the ones that have arrived at different times. Finally, it recreates the original cell sequence and forwards these cells to the ATM layer.

The major advantages of AIM

AIM offers many advantages. It provides an attractive and painless migration path between T-1 and T-3 rates. With the AIM interface, companies will have full access to ATM service classes and traffic types. This means that multimedia applications can use these levels of service and traffic options. All ATM interfaces, including UNI, NNI, and private NNI, could be used under AIM. Perhaps the major advantage is that an ATM-Forum-approved AIM specification means full interoperability among different vendors' ATM and T-1 equipment.

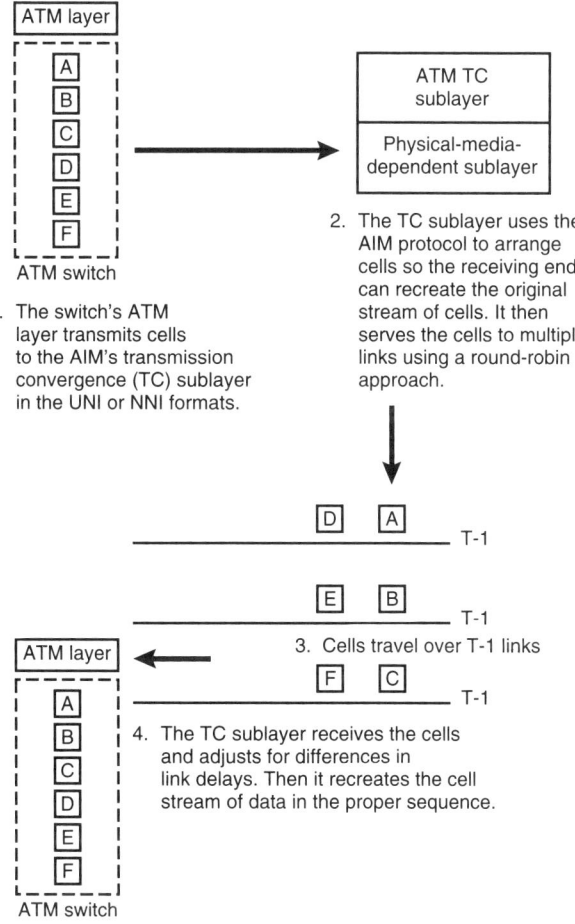

Figure 4.8 How AIM will work.

ATM and Frame Relay

Frame relay services have been growing significantly over the past few years. Service providers are looking for ways to increase bandwidth capacity, and ATM might be a solution. The connectivity between frame relay and ATM has been achieved by a set of specifications known as *frame relay/ATM PVC network interworking*, jointly developed by the Frame Relay Forum and the ATM Forum. Frame relay functions such as variable-length protocol data units, error detection, and congestion indication are supported. The specifications spell out in detail how frame relay traffic is carried over an ATM backbone network.

One of the major strengths of this interface is that it's transparent to end users. Frame relay equipment continues to transmit data to carrier's wide area network backbone switches, where it's converted from frames into ATM cells. The ATM cells travel between ATM switches and are then reconverted back into frames for transmission to frame relay equipment or routers. The beauty of this approach is that there's no need for special software in each end device to make the protocol conversion between ATM and frame relay.

There are a couple of major advantages to this now standard interface. It enables customers to connect frame relay sites seamlessly to a single or multiple sites using ATM without having to reconfigure software or hardware at any of the sites. Equally important, the interface enables customers to continue to purchase frame relay equipment knowing that it will be compatible with any ATM equipment they plan to purchase in the future. Virtually all the major carriers have indicated that they have plans to implement this interface as soon as possible.

A second set of ATM-to-frame-relay interoperability specifications is known as *frame relay ATM service internetworking*. These specifications deal with the complex subject of actually converting frame relay traffic into ATM traffic and back again. There are a number of fields that will be difficult to convert from one format to another. Traffic congestion, for example, is difficult to map across the two different technologies, and two different approaches are available under service internetworking.

The Frame User to Network Interface (FUNI)

The ATM Forum has finalized an alternative low-speed technology for ATM wide area networks. The *frame user-to-network interface (FUNI)* supports ATM data-only transmission speeds from T-1 down to 56 Kbps. The format consists of a header with two bytes, an additional two to four bytes of frame-control data, and a payload of up to 4,096 bytes. This means that the header overhead is only a fraction of 1 percent. Frame relay and ATM headers match to each other much more closely under this scheme than the service internetworking plan. All ATM signaling functions are supported, including switched virtual connections and various classes of service. ATM adaptation layers 3, 4, and 5 are supported so data segmentation and reassembly can be performed consistently by all devices on an ATM network.

FUNI frames are interoperable end-to-end with broadband UNI cells as far as signaling, traffic management, and network management. FUNI is designed to be carried as part of a higher-speed connection through standard time-division multiplexing. One practical use of FUNI's flexibility is to match appropriate speed with function. In other words, low-speed ATM traffic can share an access line with telephone circuits. A carrier's CO can take this ATM traffic and route it to a public ATM service, while the telephone circuits are routed to a virtual private network. Access lines can be shared between ATM and frame relay connections so communication costs are reduced.

Real-Life Wide Area ATM in Action

The ATM Research and Industrial Enterprise Study (ARIES) project was established by Amoco Corporation and 17 communications industry leaders, including Cisco Systems, Hewlett-Packard, and Sprint. Five of Amoco's sites were linked into an end-to-end ATM internetwork that included local area, campus, and wide area networks. The local area ATM connections ran at 100 to 155 Mbps, while the wide area network links used T-3 ATM links running at 45 Mbps. Figure 4.9 illustrates the ARIES network.

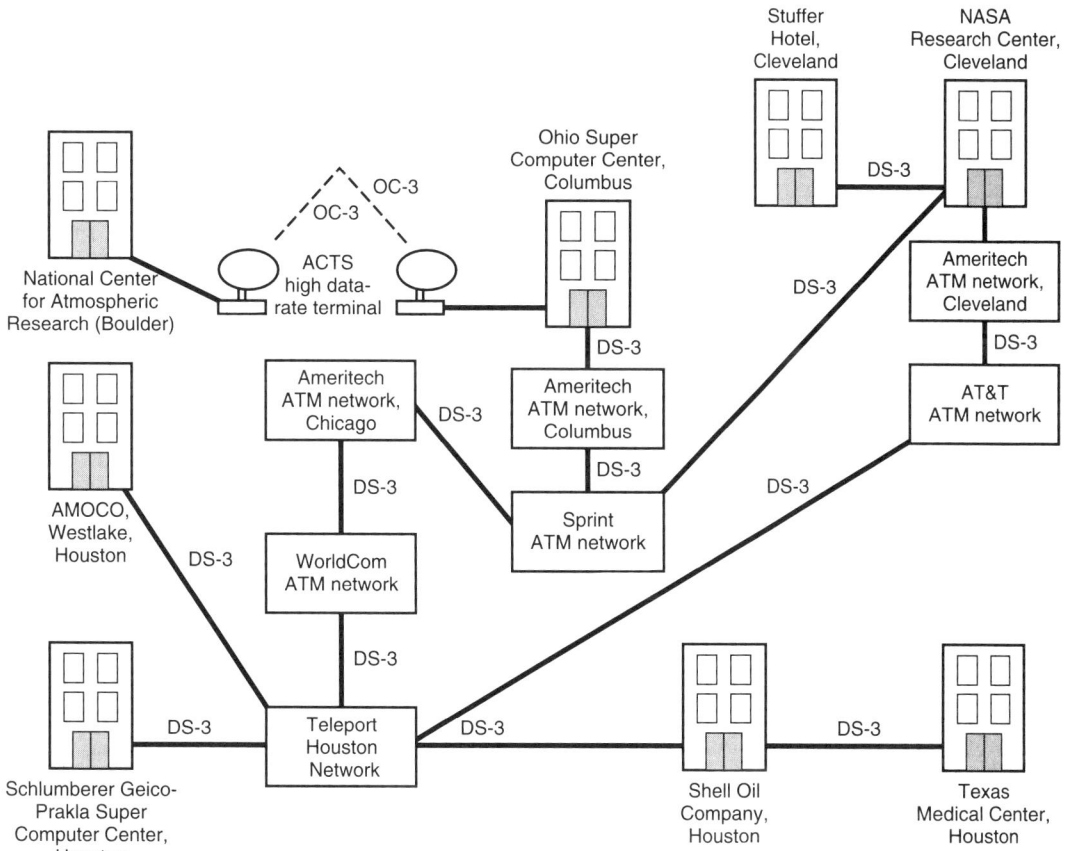

Figure 4.9 The ARIES network.

Amoco wants to develop an ATM network to help it reduce its data transmission costs. Data from remote exploration sites is transmitted to the Minnesota Supercomputer Center in Minneapolis where it's accessed from company research sites in Tulsa, Oklahoma, Houston, Texas, and Naperville, Illinois. The data is used to perform seismic modeling and simulation. In the future, Amoco would like to use its ATM network to transmit interactive video, high-resolution imagery, broadcast video, and a variety of other types of information.

ATM and Voice

One of the ways that network managers will be able to justify expensive ATM wide area network equipment is their ability to transmit voice as well as data over these links. Voice currently accounts for at least 80 percent of the traffic running over leased T-1 and T-3 lines. Voice traffic is growing at only around three percent per year, which means that the real WAN explosion in ATM usage could be fueled by the transmission of data using the savings in voice transmission costs. Now if only there were an easy way to accomplish this task.

Some ATM switch vendors such as IBM and Northern Telecom are clearly positioning their wide area ATM switches as ideal for transmission of both voice and data. Coming up with a standard way to deliver voice over ATM, however, is very difficult. Current ATM standards provide for voice traffic via *ATM adaptation layer 1 (AAL1)*. Also known as class A, this layer offers *constant bit rate (CBR)* or *circuit emulation service (CES)*. The ATM switches that currently carry AAL1 traffic do so on only a per T-1 basis. Because the entire T-1 line is treated as a single class-A virtual circuit, this approach isn't practical in the typical T-1 or T-3 private network that carries a mixture of traffic. Each different type of traffic (voice, data, and video) would require a separate T-1 access link.

Another problem with using ATM's AAL1 protocol is that its framing procedure overhead would reduce trunk capacity to transport SONET-like header information by about 8 percent, so a T-3 link would be reduced from 44.736 Mbps to 40.707 Mbps.

Any solution for carry voice and data over ATM obviously requires handling voice information more efficiently. There are several different enhancements that might achieve this goal, one of which is voice compression. Any ATM switch that supports AAL1 connections could support a voice compression device that could compress as many as eight different trunks into a single T-1 trunk. Such solutions would also have to include echo cancellation to reduce delays that would effect voice quality.

Silence suppression is another technique that could prove effective. The AAL1 protocol normally processes and transmits data coming from PBX tie lines that use channel associated signaling (CAS) to decode bits. It's possible to install a T-1 voice card that understands CAS inside an ATM switch so the switch would generate cells only when a specific voice circuit is active. Because cell generation would be reduced substantially, the transport capacity of the switch would be increased.

There's also some political considerations for companies that consider ATM switches for private ATM networks and need to be able to transmit both voice and data to cost-justify these switches. As data communications and telecommunications have begun growing more closely together, there has been a noticeable trend over the past few years toward consolidation of power and budget authority on the IS side. Still, in many companies the budgets remain separate. This means that telecommunications managers are likely to have some reservations about draining their budgets to benefit the data communications managers' need for greater bandwidth in wide area data transmission.

ATM and Wide Area Network Switches

In chapter 10, you'll look at a number of ATM switches offered by wide area network ATM vendors. They fall into two basic categories: core switches and edge switches. The carrier *core switches* are positioned for central offices. They provide the brute power required to move huge amounts of traffic that can sometimes exceed 100 Gbps. The real intelligence on the wide area network, however, resides in the *edge switches*, such as General DataComm's APEX and Cascade's 500 products.

Edge switches generally concentrate customer traffic and transmit it to a carrier's backbone. Cascade, for example, believes that its edge switches are ideal residing at the edge of the public network where they can act as traffic cops to concentrate customers' premise traffic and direct it. Other vendors, such as Newbridge, have begun to position its edge switches as premises equipment located at a customer's site where it can receive all kinds of direct traffic, including video, LAN, frame relay, and circuit emulation traffic. Unlike core switches, edge switches need a variety of different speed interfaces to handle this diverse traffic.

Carrier Implementations of ATM

Wide area network ATM specifications haven't progressed nearly as quickly as local area network specifications. Still, several vendors are offering ATM service. You can start with your local telephone company. If your network crosses a local access and transport area (LATA), then you'll also need to talk with an interexchange (IXC) or long distance carrier such as AT&T, MCI, or Sprint. This group can provide end-to-end ATM service in conjunction with your local carrier, which provides the access lines between your premises and the long distance company's point of presence (POP) or central switching office. There are also bypass providers such as MFS, Datanet, and Teleport, which can provide ATM service.

The ATM services being offered differ widely. Some carriers offer a flat monthly rate for a port connection based on speed. There's also usually a flat fee per virtual connection based on speed. There might be additional charges for distance and usage. Some offer constant bit rate (CBR) while others offer variable bit rate (VBR) service or available bit rate (ABR). These options were described in detail in chapter 2. Because the current ATM wide area service providers mix standards-based with proprietary-based hardware and software, it's difficult at present to build private ATM networks that mix and match different switches.

MFS Datanet offers one of the first integrated voice/data ATM services. The *wide area voice exchange (WAVE)* enables the company's customers of its private ATM network to send voice and video traffic from site to site. Users connect their LANs and PBX tieline trunks to an MFS Datanet-operated data service unit (DSU). Voice

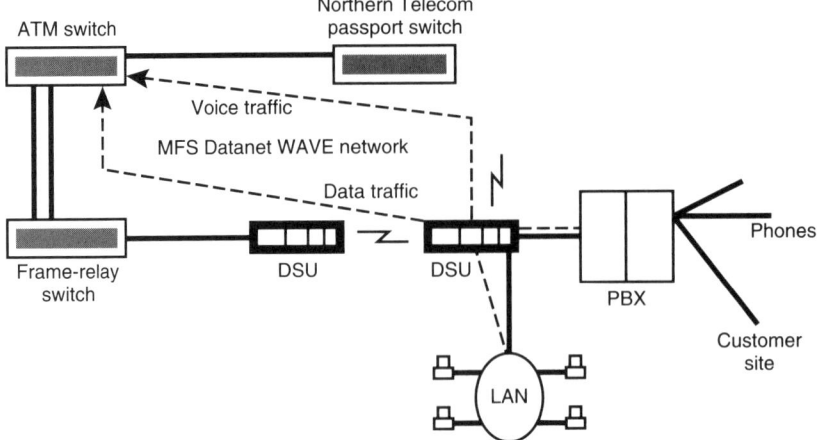

Figure 4.10 The WAVE system in action.

traffic is routed to a Northern Telecom Magellan Passport switch at MFS Datanet's San Jose, California, headquarters. The ATM switch determines which channels are congested and reroutes voice traffic around them. It uses a prioritization scheme to give voice cells precedence over data cells. MFS Datanet believes that consolidating voice and data traffic over its network can save customers 10 to 15 percent over comparable data and voice WAN networks that use frame relay and public leased lines. WAVE began in Chicago, Los Angeles, New York, San Francisco, and San Jose. The company has plans for 16 cities before the end of 1996. Figure 4.10 illustrates the WAVE system in action.

With ATM services still in their infancy, customers must be sure to check for any possible interoperability problems because not all specifications are currently in place. Prospective WAVE customers, for example, must be aware that there are no industry standards for the type of voice service MFS Datanet offers and should check with equipment vendors to ensure that they use a common voice encoding technology.

Case Study: Using the Public ATM Network Today

There are some major advantages to living in North Carolina besides the relatively low cost of living. One such advantage is the state's investment in the information highway, namely, the North Carolina Information Highway's (NCIH) public *asynchronous transfer mode network*. Developed in conjunction with North Carolina state government as well as local exchange carriers, its purpose was initially to provide ATM links for public sites. Bell South is now actively soliciting commercial business.

Branch Banking & Trust Company is one such business taking advantage of this public ATM network. One of the major strengths of public ATM is its ability to reroute data traffic easily to accommodate moves and changes. With 452 banking offices in 220 cities in three states, the bank is attempting to consolidate buildings by shifting people around. ATM makes the changes economical.

Figure 4.11 shows that the bank is currently using public ATM to connect LANs in three buildings. It uses left-over bandwidth to interconnect telephone networks. While prices will likely have changed by the time you read this book, the bank discovered that ATM costing $7540 per month was 20% less expensive than a DS3 option, factoring in the cost of three FiberCom 7650 ATM cell edge multiplexers amortized over three years.

Many of the bank's LAN segments feed into Proteon routers with 45-Mbps DS-3 data service units connecting to edge multiplexers. Each of the three buildings has an OC-3 connection over a Bell South SONET ring going to a Fujitsu ATM central office switch.

The bank's OC-3 lines going to each remote office are sent into a DS3 ATM line through the Fujitsu ATM switch. The remaining bandwidth is used by the branch offices to carve out T-1 PVCs to splice together the voice networks in the remote offices. Because OC-3 service is already installed, the bank was able to add two T-1 PVC connections costing $420 per month, compared to the $800 per month cost of one stand-alone T-1 connection. In the near future, the bank plans to add a fourth site to this network that will be tied in by a DS-3 link to one of the remote offices. This new site will use routers to link up with the CO ATM switch and then to headquarters using the left-over bandwidth on the OC-3 line.

Summary

Digital lines that offer 1.544-Mbps bandwidth (T-1) or 45-Mbps bandwidth (T-3) have proven to be enormously popular. They are cost-effective and can provide wide area network redundancy. This bandwidth is generated using the technique of time-division multiplexing in which data streams are allocated specific time slots within frames that are being transmitted. For companies with heavy traffic, these digital lines are economical.

For wide area networks, the X.25 protocol is still widely used. It uses a technique known as packet switching. While it has a maximum transmission speed of only 64 Kbps, it also has excellent error checking. Because users are generally charged for

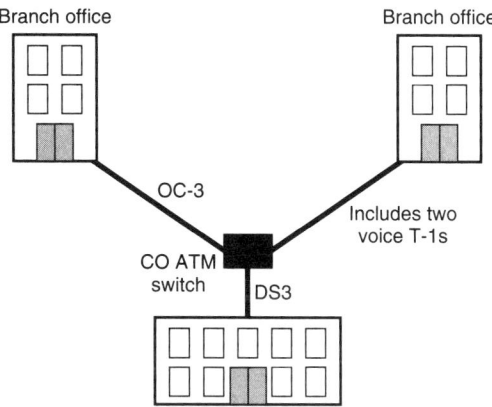

Figure 4.11 A bank implements a wide area network public ATM.

the number of packets they send and not the distance, this technology is attractive for global network users. Frame relay is a newer packet-switched approach that features variable-length packets and greater efficiency, which permits up to a 2-Mbps transmission speed.

A competitor to frame relay is switched multimegabit data service (SMDS). This connectionless service can deliver transmission speeds of up 45 Mbps. It's based on the IEEE 802.6 protocol developed for transmission of data over a metropolitan area network. Because its cell structure is very similar to that of ATM, SMDS networks will most likely become part of ATM networks without much trouble.

For transmission of voice, data, and video information, integrated services digital network (ISDN) is becoming a viable option. A basic rate interface (BRI) offers a single access point into an ISDN network. It provides two bearer channels of 64 Kbps and one data channel of 16 Kbps, so it's also known as 2B+D. For companies that want multiple access to an ISDN network for devices such as a PBX or LAN, there's a primary rate interface (PRI), also known as 23B+2D because it offers 23 bearer channels of 64 Kbps each, and two data channels of 64 Kbps each.

There's considerable work taking place in linking slower-speed T-1 and frame relay networks to ATM networks so network managers can consolidate traffic and benefit from ATM's speed and services. Asynchronous transfer mode inverse multiplexing (AIM) is a technique that permits T-1 traffic to use ATM links. The ATM Forum is also developing an interface between frame relay and ATM. One controversy involves whether frame relay users would have to purchase new equipment to be able to use such an interface.

One key issue for wide area networks is whether or not ATM switches will be able to carry voice as well as data information efficiently and economically. There are several problems that must be resolved to achieve this goal, including the efficient compression of voice information and how this information will be transmitted via ATM.

Vendor Network Architectures and ATM Plans

FORE Systems Leads the Way

FORE Systems has been a leader in the ATM field from the very beginning. Dedicated solely to ATM technology, this startup company quickly established itself as the market leader. It placed its first ForeRunner ATM switches in workgroups that formed experimental test beds in several major U.S. corporations. Today, it has a well-established family of ATM workgroup and backbone switches. In addition, it has consistently supplemented ATM Forum standards with proprietary algorithms to fill in gaps. In this chapter, I'll describe FORE Systems' hardware and software products, including those for network management and traffic control.

FORE Systems' Strategy

FORE was out the gate quickly in the ATM marketplace. With very few standards in place, the company developed its own proprietary algorithms and established itself as the market leader. By 1995 it controlled between 65 and 70 percent of the still-emerging ATM market. When LAN emulation was far away from an ATM Forum standard, FORE customers tended to use the equipment for workgroups composed of high-performance workstations. Those who wanted to test LAN emulation did so with FORE's own proprietary hardware and software. Sharp price drops in 1995 brought the price of ATM for the PC desktop to a level that began to attract commercial customers. As the installed base of ATM workgroup switches has grown, FORE has expanded its product line to include backbone switches so ATM networks can be formed. While including numerous wide area network interfaces, it has chosen not to develop switches for the WAN core switch/central office market.

FORE had been at a disadvantage, however, because of its lack of Ethernet switches. It required customers to commit completely to ATM while rivals in some cases offered switches that could be migrated to handle ATM cells. This situation changed when FORE acquired the Ethernet switching companies, Applied Network Technology, and

Alantec. Owning the technology is important to FORE for another competitive reason. Given the price sensitivity of the Ethernet switching market, an OEM arrangement wouldn't enable FORE to offer products that are price-competitive.

FORE is now able to offer Applied Network's desktop Ethernet switch; its own Ethernet, Token-Ring, and FDDI-to-ATM switch; and a third product, an Ethernet switch that's a hybrid of the other two products at a much lower per-port price.

A key part of FORE's strategy has been to build the ATM market by lowering prices. It needs to establish a large enough installed base of users to withstand the assault by much larger companies with deeper pockets, such as IBM and Digital Equipment Corporation. Because of its need to commit significant revenue toward continued research and development, FORE's profit margin has lagged behind what's normally found when technology is still in its infancy. Despite this situation, FORE's stock value has soared, a testimony to its technological leadership position within the ATM industry. The key question is whether or not FORE can continue to develop and sell ATM products with lowered profit margins and increased competition. One obvious solution is FORE's partnership with vendors that can seed its technology in markets that it cannot cover itself. This concept is realized in the ForeThought Partners Program.

The ForeThought Partners Program

Clearly FORE Systems isn't big enough to reach all potential customers and fortify its grip on the ATM market. The ForeThought program is a way to increase revenue and market penetration, and cement strategic relationships with other companies. The program makes the ForeThought software available to a variety of other programs. By the close of fiscal year 1995, 35 partners had joined the program. Among the companies that have joined the program are Cabletron, Northern Telecom, Optical Data Systems, and Tricord.

The Northern Telecom partnership is particularly important because of this company's leadership role in the wide area network market. I'll provide a closer look at how FORE and Northern Telecom will work together a bit later in this chapter when I focus on FORE's wide area network interfaces. FORE and LANNET (now owned by Madge Networks) have an arrangement that enables LANNET to integrate FORE's ATM technology into its MultiNet switching hub line of products. The integration takes the form of an ATM edge adapter module that will permit transmission of switched packets from the MultiNet's 1.28-Gbps backplane over an ATM backbone. LANNET will also make use of the ForeThought internetworking software.

FORE has a similar arrangement with another hub vendor, ALANTEC. ALANTEC is porting FORE's entire suite of ATM access software to its PowerHub family of switching hubs so the hubs can connect switched Ethernet, switched FDDI LANs, and ATM networks.

FORE's Switches

FORE refers to its switches as "second generation," meaning they're designed for the commercial market and not simply for customers who want to test ATM. FORE's first generation of switches tended to go to research universities and entertainment-oriented companies. It feels that the ATM market is now ready for true commercial applications. The switches use a nonblocking architecture, meaning that the switching fabric's capacity is greater than or equal to the sum of a switch's individual input/output port capacities. Each switch contains its own RISC processors, and FORE stresses that the distributed environment of FORE switches increases fault tolerance as well as performance because it eliminates a single point of failure. The switches are modular and can be configured for between two and 96 ports.

Backbone switches

FORE differentiates between its backbone switches and its workgroup switches. The ForeRunner ASX-200 BX/BXE backbone switches are designed specifically for backbone use. They can connect up to 24 LAN servers or LAN access devices such as hubs, routers, and LAN switches. The switches are expandable to 10 Gbps of bandwidth. Port speeds vary from 1.5 Mbps to 622 Mbps with support for virtually all major LAN and WAN interfaces. Because they're designed for the backbone, FORE has added certain key features such as bandwidth management and on-demand switched virtual circuits, and virtually all components, including power supplies, fans, network modules, CPUs, and switch fabrics, are hot-swappable. Each 2.5-Gbps switching fabric houses a separate CPU. In the event of a single CPU or single switch fabric failure, the remainder of the switch is unaffected and continues to operate.

FORE is positioning itself as a major player in the backbone market with its ASX1000. This nonblocking switch is scalable to 10 Gbps of bandwidth. It supports up to 96 ports, including virtually every major LAN and WAN interface, as well as speeds up to 2.5 Gbps. Because of its role as a central backbone switch, FORE has added an I960 integrated processor for each of its switching fabrics, which can be hot-swapped. It has also increased the buffer size for this switch by a factor of four over its other switches to 212K cells.

Workgroup switches

The ForeRunner ASX-200WG is FORE's workgroup switch. It contains the same architecture as the backbone ASX-200BX model and some of the same features. It's a 2.5-Gbps nonblocking switch that uses I960 processors for distributed intelligence and offers LAN interfaces from 25 to 622 Mbps. It contains the same reliable features found on the backbone models (hot-swappable network modules and redundant fans). The central differences (besides the cost) are the workgroup-size configurations (12 to 24 preconfigured ports) and the lack of wide area network interfaces. Figure 5.1 illustrates a FORE-Systems-configured ATM network that includes both an ATM backbone switch and ATM workgroup switches.

Ethernet switches

FORE offers the ForeRunner ES-3810, the first product in a family of ATM-ready Ethernet switches. These switches are interoperable with Alantec's switching hubs. FORE's Ethernet switch offers remote monitoring support on each LAN segment or

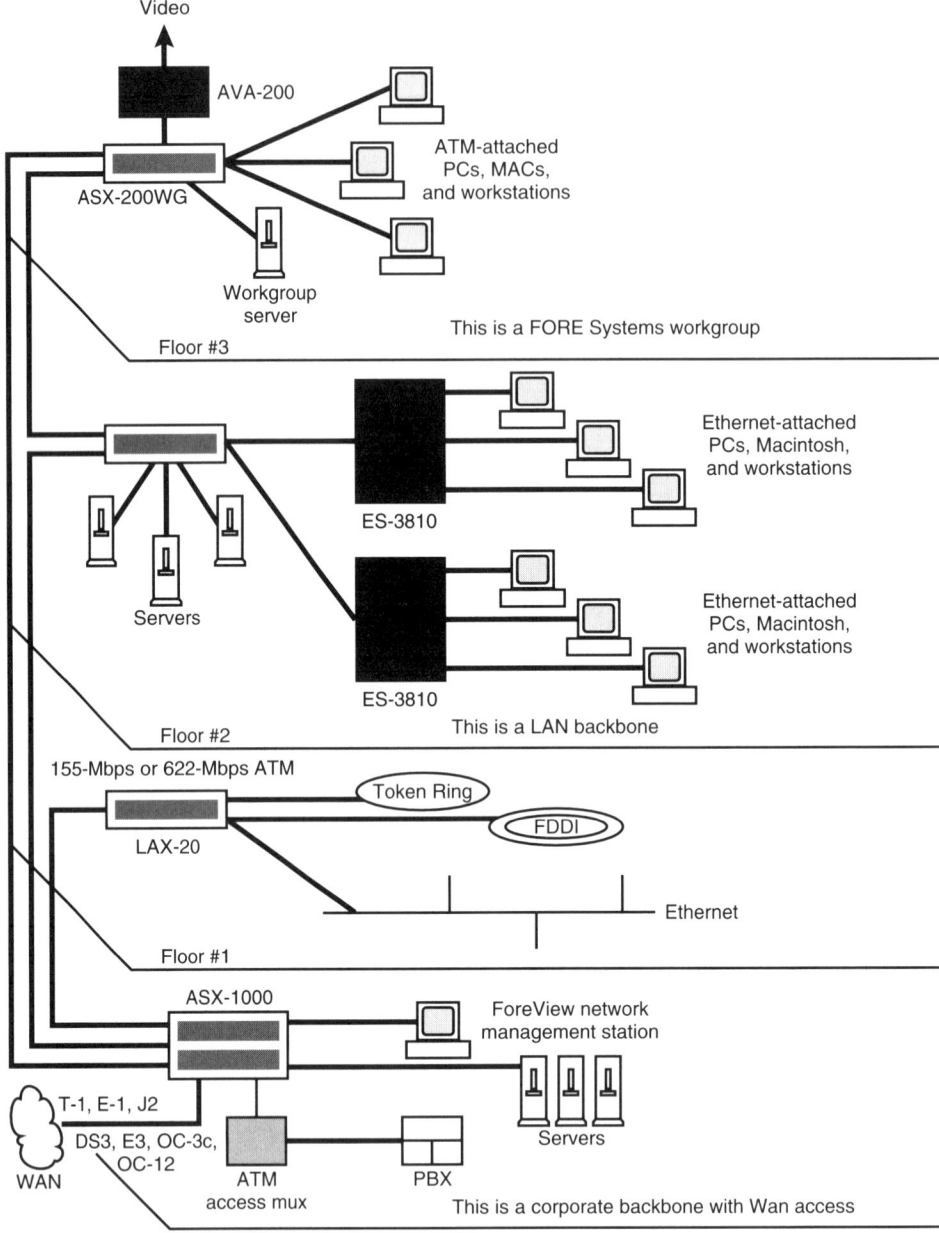

Figure 5.1 A FORE Systems ATM network with backbone and workgroup switches.

port and supports ForeView network management software. The switch will also support ATM modules. This promise of a migration path makes it particularly attractive for network managers who want to ease into ATM.

The LAX-20 LAN access switch

The networking world is still a world of legacy LANs, so routing between that world and the new world of ATM is crucial. The LAX-20 LAN Access switch provides internetworking between legacy LANs (Ethernet, Token Ring, and FDDI) and ATM networks. It also provides intelligent local switching of LAN traffic for up to 16 dedicated 10-Mbps Ethernets, 40 semiprivate Ethernets, 12 dedicated Token Rings, or multiple FDDI rings. It comes with SNMP-based network management as well as the ForeThought software.

Perhaps the easiest way to view the LAX-20 is as a router with slots for ATM, Ethernet, Token Ring, and FDDI modules. It uses FORE's LAN emulation software to translate between ATM and LAN topologies.

A FORE ATM Network in Action

One of the reasons it's now possible to write a book on the second, commercial phase of ATM is that there are real and not merely prototype or experimental ATM networks being used in several different industries. One example is an ATM network with LAN emulation found at the Wall Street institutional equities house of Donaldson, Lufkin, & Jenrette (DLJ), one of the first financial companies to use ATM technology on desktop to support live stock trading.

DLJ's Repurchase Group is one of the company's divisions that handles the buying and selling of U.S. Government securities. Originally it used Intel 80386-based IBM PCs attached to a server on a single Ethernet LAN segment. This network evolved into a two-segment FDDI network of 35 IBM PS/2 computers running DOS and attached to a server running four custom real-time trading system applications under the NetWare 3.11 network operating system (NOS). Traffic has continued to grow, so the group looked to ATM as a way of increasing bandwidth and thus response time.

Each client and server is now equipped with ForeRunner ESA-200PC ATM adapter cords. Multimode fiber-optic cabling links these PCs to three ASX-200 ATM switches. Intel 40846-based PCs are linked via 100-Mbps TAXI links, while Pentium-based PCs use 155-Mbps SONET links. The ATM network is managed from a Sun SPARCstation running ForeView network management software. (This software will be discussed a bit later in the chapter.) DLJ believes it has achieved a 50% reduction in the time required to process triparty trades. Figure 5.2 illustrates the DLJ ATM network.

There are several other applications that FORE Systems believes will drive the sales of its ATM switches and software. Cinebase by Visual f/x Inc. is the only resolution-independent relational database designed specifically for visual asset management. Format- and network-independent, it can manage virtually every form of digital media in any environment. It can navigate though volumes of motion pictures, animation, video clips, photographs, medical images, 3-D geometry, text, and audio. FORE has included this application in its industry convention application showcase.

FORE's ATM Adapters

The ForeRunner ATM adapters contain four key elements: an enhanced SAR processor (the ESP chip), several FORE-developed bus-specific ASICs, an I960 RISC microprocessor, and an integrated physical-media-dependent (PMD) interface. The

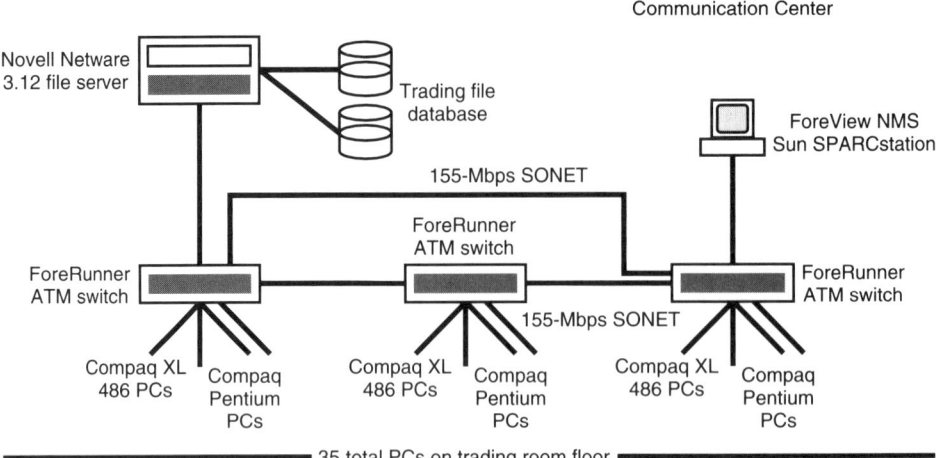

Figure 5.2 The Donaldson, Lufkin, & Jennette ATM network. Note that 486 PCs are connected via 100-Mbps TAXI, and Pentium PCs via 155-Mbps SONET links.

ESP chip provides the hardware-assisted ABR that will eventually be supported in ForeThought software. FORE-designed application-specific integrated circuits (ASICs) are tuned for each specific host bus. The I960 RISC processor is an onboard engine that simplified software/firmware upgrades. The integrated PMD components are integrated on the adapter card for cost and reliability reasons. Figure 5.3 illustrates a typical FORE ATM adapter card.

In addition to support for PCs, FORE also supports the Macintosh platform with its NuBus adapter cards. These cards are scalable from 25 Mbps to 155 Mbps. They contain a 25-MHz I960 RISC processor. All cards include Macintosh drivers for AppleTalk and MacTCP and are compatible with FORE's ForeThought software; thus they're also compatible with the ATM Forum's specifications for LAN emulation. FORE also plans to offer peripheral component interconnect (PCI) cards for Power Macintosh models because NuBus doesn't offer the performance customers need. Power Macintosh computers with PCI buses and Apple's open transport reworked networking architectures should provide close to wire speed, minus ATM overhead.

ForeThought Software

ForeThought is the company's suite of software components. It supports all current ATM standards, including the user-to-network interface (UNI) 3.0 Q.2931 for switched virtual-circuit signaling. This means that current specifications for LAN emulation, TCP/IP over ATM, and congestion management are also part of this software. FORE has added its own proprietary *simple protocol for ATM network signaling (SPANS)*, which includes NNI signaling. The addition of SPANS means that customers can take advantage of such advanced networking features as IP multicasting, automatic network configuration, intelligent optimized routing, link load balancing, alternate routing, and SVC tunneling. SPANS permits the creation of switched virtual circuits between switches.

IP multicasting mimics the operation of IP over conventional LANs. It's performed in the switch nearest the recipients rather than at the source to increase network efficiency. Figure 5.4 shows the relationship between ForeThought software and FORE's internetworking architecture.

Figure 5.3 FORE Systems ATM adapter cards. (*FORE Systems*)

ForeThought internetworking software	
LAN emulation	Virtual workgroups
IP over ATM	SVCs

ATM switches	LAN access switches	ATM adapters
ForeRunner Hardware		

ForeView NMS
Virtual WG administration
Mapping
Configuration
Security
Monitoring
Fault management

Figure 5.4 FORE's internetwork architecture.

ForeThought and LAN emulation

FORE is committed to having its ForeThought software fully compliant with the latest LAN emulation standards from the ATM Forum. Supporting the current set of ATM specifications means that ForeThought provides a LAN emulation configuration service (LECS). All optional configuration parameters defined in the LECS to the LAN emulation client (LEC) protocol are also supported. The LAN emulation server (LES) provides an address resolution (ARP) service. Clients send address queries to the LES and receive responses from the LES or from other clients. A broadcast and unknown server to handle special-case traffic is also supported.

It's possible to have multiple, separate emulated LANs. Each emulated LAN acts like a distinct LAN with respect to any other emulated LAN, and each LAN has its own bus so broadcast and unknown traffic is seen only by the members of that particular emulated LAN. FORE has indicated that it plans to support emulated FDDI LANs.

ForeView

ForeView is an SNMP-based, graphical-oriented management system. Users can configure, control, monitor, and troubleshoot ATM networks. It's possible to configure smart permanent virtual circuits (SPVCs), which provide automatic configuration and rerouting of PVCs in the event of network disruption. ForeView integrates with other network management platforms such as HP OpenView, SunNet Manager, and NetView/6000. It runs as an element under OpenView and SunNet Manager. ForeView supports Windows NT, Solaris, SunOS, and HP-UX operating systems. The software will run on a Sun SPARCstation, HP-700, or 486/66 PC with at least 32MB of memory and 20MB of free disk space.

The inventory management component of this software keeps track of all FORE ATM hardware and software deployed across a network. The result is automatic discovery and mapping of FORE's ATM networks. It compiles information relating to hardware and software revisions, serial numbers, and addresses and node names. The software makes it possible to track network usage on a per-host or per-link basis. It offers in-band management over ATM and out-of-band management over Ethernet.

You can also use a scripting utility to handle moves, adds, and changes. There's also an online reference manual that provides context-sensitive text and diagrams. Figure 5.5 illustrates a ForeThought screen.

Support for IP over ATM

FORE switches support ATM-based services for the transparent support of the TCP/IP protocol suite based on the IETF RFC 1483 and 1577 standards, as well as IP multicast.

Customer Support and Product Maintenance

FORE offers its own support as well as distributor support. In 1995 it had 11 offices in the U.S. and 17 offices worldwide. Its ForeMAN support programs provide com-

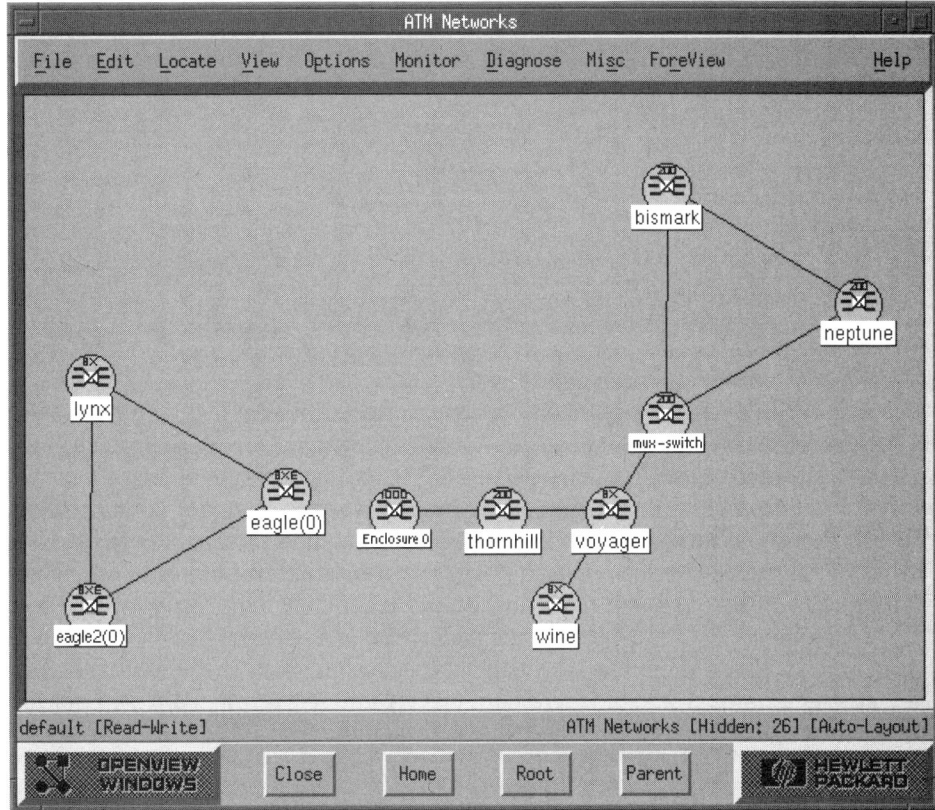

Figure 5.5 A ForeThought network management screen. (*FORE Systems*)

prehensive support to customers. This program includes software enhancements, software corrections, and hardware replacement. Technical support is offered via phone, fax, or e-mail. Unlimited e-mail support is available 24 hours a day, seven days a week. Response to e-mail is guaranteed within one business day. There's a one-year warranty on hardware.

Traffic Policing

FORE switch support adheres to the ATM Forum's standard for traffic policing known as "dual leaky buckets." One bucket monitors the average bandwidth while the other one monitors peak bandwidth. By policing traffic congestion, the leaky bucket technique enables carriers to offer bandwidth on a pay-as-you-go basis to the commercial market so customers don't have to pay for unused bandwidth. These companies are able to contract with carriers to pay based on average and peak bandwidth usage. (This approach was discussed in chapter 2.) When bandwidth usage exceeds what's requested, customers can choose to have the cells either automatically discarded from the network or tagged as being in excess of the allocated bandwidth and allowed to pass through the network as long as available bandwidth exists.

ForeThought Bandwidth Management

ForeThought bandwidth management offers a number of features, including per-VC queuing, smart buffers, and packet-level discard. Because these terms are quite technical, let's take a few minutes looking at what the features actually provide. The *per-VC (virtual circuit) queuing* feature is a buffer-management technique that provides a dedicated queue for each virtual circuit. Each network module can support up to 12,000 virtual circuits. The advantage of this feature is that it enables an ATM switch to perform congestion avoidance techniques to guarantee multiple service levels and packet level discard (to be discussed shortly). Because each connection is treated independently, there's no chance of high-priority traffic being forced to queue behind lower-priority traffic. This particular problem is sometimes referred to as "head-of-the-line blocking."

Smart buffers manage bandwidth efficiently and maintain the quality of service for each connection. The added efficiency comes from their ability to dynamically and automatically allocate space based on a connection's service-level requirements. FORE believes that this feature increases the effective size of buffers by 1½ to 3 times to provide an effective capacity of 75,000 to 150,000 cells per switch. These buffers are "smart" because they can prioritize traffic into multiple service levels. They have a maximum capacity of 13,312 cells and serve as a cushion when the amount of traffic exceeds the bandwidth of an outgoing port.

Smart buffers can be particularly valuable when they're located at the edge of a wide area network where data travels from a high-speed LAN port such as a 155-Mbps SONET to a much slower WAN port such as a 45-Mbps DS3, 1.5-Mbps DS1, 34-Mbps E3, or 2-Mbps E1.

These buffers offer 381 service priorities. Each output port has three service levels, and each service level has 127 sublevels. Because there are three service levels, there's a guarantee of a low cell delay variation (CDV) for CBR, medium cell delay variation for VBR, and unspecified cell delay variation for ABR and UBR. They can be automatically configured or manually configured by the user. The allocation scheme for smart buffers ensures that ABR and UBR traffic won't infringe on the CBR and VBR allocated buffer space, so CBR and VBR traffic is protected from ABR and UBR traffic.

AAL5 packets now account for more than 95% of traffic on a typical ATM network. An AAL5 packet consists of several cells. If even a single cell of that packet is dropped because of link congestion, the entire packet is corrupted. This bad packet would normally go through the network and waste valuable buffer space and occupy valuable bandwidth. When the data has to be rebroadcast, the problem is compounded because additional bandwidth is required, congestion is increased, and the possibility that there will be additional corrupted packets is increased. FORE has therefore designed *packet-level discard* to deal with this problem. Packet-level discard discards groups of cells that form entire packets rather than discarding random cells from multiple packets. FORE believes that by dropping a small number of complete packets instead of random cells from a large number of packets, "goodput" can be increased by a factor of two to five to make the network more efficient.

FORE switches also offer early packet discard (EPD) and partial packet discard (PPD). *Early packet discard (EPD)* is used if the total number of cells in the output buffer exceeds a user-established threshold. EPD prevents new packets from entering the buffers where they would undoubtedly lose cells because of congestion. EPD takes the remaining buffer space and reserves it for packets that have already begun to enter the buffer; this increases the chances that these packets will be successfully transmitted. Once the cells in the buffers fall below the threshold, EPD permits packets to enter the buffer.

Partial Packet Discard (PPD) handles the situation where buffers overflow despite EPD and a cell must be discarded. PPD discards all remaining cells in the packet (the tail of the packet) that has already started entering the buffer. FORE points out that PPD works closely with EPD and clusters the discarded cells into the fewest number of packet tails to reduce the overall number of packet tails dropped. The result is that the buffer level quickly falls below the EPD threshold.

The Role of 100-Mbps ATM

FORE ships an ASX-200WG switch with a configuration of either 10-port/12-port 155-Mbps ATM or 18-port/24-port 100-Mbps ATM. There are also interfaces for 622-Mbps ATM. FORE's price reductions make the 100-Mbps ports competitive with FDDI, attracting customers who still have traditional 10-Mbps Ethernet backbones and know that they have to move to a faster system.

FORE and 25-Mbps Desktop ATM

FORE has not been an advocate of 25-Mbps ATM, but it's now a standard and the market is building for it. The result is that the company reluctantly issued a statement that says, in effect, that there are two situations when it can imagine customers using 25-Mbps ATM:

■ When the network is used to interconnect computers whose bus would be saturated by a 155-Mbps network connection (e.g., older models of computers)

■ When the network's existing physical wiring plant consists mainly of category-3 copper wiring that can't support 155-Mbps transmission speeds

With the addition of 25-Mbps ATM, FORE's interface options are as follows:

Speed	Media
155-Mbps OC-3c/STM-1	Multimode fiber
155-Mbps STS-1/STM-1	UTP category 5
100-Mbps TAXI	Multimode fiber
25 Mbps	UTP category 3, UTP category 5

ForeRunner Video Products

One of the early drivers of ATM was video applications because of the high-bandwidth requirements for high-quality audio and video. FORE has developed the Fore-Runner AVA-200 ATM video adapter. It multicasts real-time, high-quality video and audio from video sources to ATM-connected workstations and PCs. The video adapter in conjunction with the ForeRunner AVD-200 provides full-motion, real-time video and CD-quality audio over an ATM network on standard monitors, televisions, and large-screen displays.

The AVD 200 can receive a video stream off the ATM network and put it onto a full-screen monitor at 30 frames per second. It delivers CD-quality audio at 1 Mbps and video at 3.5 Mbps. The National Aeronautics and Space Administration (NASA) put on a demonstration of one-to-many desktop video conferencing. The Ames Research Center at Moffet Field, California, joined with the Langley Research Center in Hampton, Virginia, and the Lewis Research Center in Cleveland, Ohio. Each site was equipped with an AVA-200 ATM video adapter connected to an ATM switch from FORE Systems. The interface between the two was a 100-Mbps TAXI connection. AVA-200s and video cameras were set up at each site. Video from all three sites were displayed on a single workstation at each location. There was virtually no delay across an ATM connection of more than 10,000 miles. During the demonstration, uncompressed video was sent over ATM permanent virtual circuits at 24 frames per second.

FORE's Vision of Enterprise ATM

While FORE Systems began as a workgroup switch company, it's now eyeing the backbone as its next logical evolution. Its vision of the future enterprise network places routers at the periphery of a switched network. It has announced that it will offer a series of services as part of a layered architecture it calls *ForeThought internetwork architecture*. Here are the major services provided by each of the four key layers:

Layer 1: ATM transport services. Serving as the foundation for the entire switching model, this layer's services convert all non-ATM traffic to ATM cells at the edge of the network. It includes support for connectionless and connection-oriented services over a connection-oriented ATM backbone. The layer supports Fore's switches and adapters as well as its LAN and WAN access devices.

Layer 2: virtual LAN services. This layer defines a VLAN as a logical association of users that share a common broadcast domain. Included in this layer is support for the ATM Forum's LAN emulation specifications as well as the IETF's support for IP over ATM.

Layer 3: distributed routing services. This layer transforms today's centralized routers into tomorrow's distributed routing functionality. Services support communications between virtual LANs, conversion between different MAC types such as Token Ring and Ethernet, and scaling to global networks.

Layer 4: application services. This layer provides services for access control, security, connection aiding, virtual-circuit management, and quality of service based on bandwidth reservation.

As part of FORE's plan, ForeView management software will extend to provide network management of all the services embedded in the ForeThought ATM internetwork architecture.

FORE's Vision of ATM and the Wide Area Network Market

Systems sees the evolution of ATM wide area networks proceeding rapidly. In 1993, wide area ATM consisted of interconnecting LANs over private lines. By 1994, production LANs were connected over pilot wide area network ATM service networks using switched virtual-circuit (SVC) tunneling or private virtual circuits across carrier ATM permanent virtual-circuit services. FORE believes that by 1996 true integrated data, image, voice and video ATM networks will be a reality, with seamless LAN-to-WAN integration.

Northern Telecom's Strategic Alliance with FORE Systems

One real problem for FORE Systems in selling to Fortune 1000 companies has been its size. Because specifications for wide area network ATM service lags behind LAN specifications for ATM, many customers who want to be able to eventually link their ATM LANs and WANs in a seamless manner are reluctant to go with a company that might have only part of an ATM solution. FORE Systems' real strength has been in its workgroup switches. It isn't inclined nor does it have the resources to play in the wide area ATM switch market. Because LAN-to-WAN integration via ATM is likely to involve proprietary algorithms for quite a while, customers should think twice about trying to mix and match FORE's switches with those of wide area network vendors.

FORE has solved this problem by forging a strategic alliance with Northern Telecom. FORE's ForeThought software will be added to Northern Telecom's Magellan Passport and Concorde switches, effectively providing seamless LAN-to-WAN integration via the same ATM software. This means that the Northern Telecom switches will include FORE's functionality with switched virtual circuits, IP multicasting, LAN emulation, automatic network configuration, and intelligent optimized routing. Equally important, elements of FORE's ForeView network management software will be incorporated into Northern Telecom's carrier network management system. The companies will also participate in joint development projects.

This is one of those industry relationships that's good for both players. You've already seen what FORE gets from this partnership. Northern Telecom's sales force gains entry to FORE's customer base. Now they can offer these people guaranteed interoperability between LANs and WANs.

FORE's Wide Area Network ATM Interfaces

FORE offers both low-bandwidth and high-bandwidth WAN interfaces for its ATM products for connectivity with new-carrier ATM services and existing-carrier private lines. At the low end, it offers T-1 (1.544 Mbps) for the North American market, E-1 (2.048 Mbps) for the European market, and JT2 (6.312 Mbps) for the Japanese market. At the high end, interfaces include DS-3 (45 Mbps), E-3 (34 Mbps), and OC-3c/STM-1 (155 Mbps).

FORE believes that the T-1 interface will increase demand for ATM services because of the large amount of T-1 lines installed. It sees the T-1 wide area network interface as ideal for linking remote sites to ATM backbones.

Should FORE Systems Be Your ATM Vendor?

FORE Systems offers a number of attractive reasons to make it your ATM vendor of choice. It has been the dominant ATM market leader from the beginning and has been a leading voice within the ATM Forum. It has been among those pushing for ATM Forum specifications for remote network management and other key features. No industry critics question FORE's technology; it has remained leading-edge. The addition of its own Ethernet switching products makes it a better choice for companies looking for a vendor that can migrate it slowly into an ATM network beginning with Ethernet switching. FORE has made it clear that it will establish migration paths for users so they don't feel trapped with dead-end products.

The strategic partnership with Northern Telecom is crucial for FORE because it allows it to offer the seamless LAN-to-WAN interoperability over an ATM network. In fact, it's a very compelling argument for companies to look at a one-vendor FORE Systems solution if they can't wait for the ATM Forum to fill in the serious gaps that still remain in wide area network ATM specifications.

FORE's commitment to supporting Apple products might make it particularly attractive to companies with mixed environments and a significant number of Macintosh and Power PC computers that need to be integrated into an ATM network.

While FORE offers many advantages as an ATM vendor, some customers might not match up well with FORE's strengths. At present, FORE doesn't offer Token Ring switching. While this situation might change and FORE might resell a product manufactured by another vendor, customers with a large installed base of Token Ring equipment might feel more comfortable with a different vendor, such as IBM, which is more versed in this technology. This decision is particularly appropriate if the customer feels more comfortable with a one-vendor solution to its enterprise network needs.

Customer support is another area where FORE Systems is bound to be attacked by its competitors. While its support seems perfectly fine at this time, companies such as IBM and Digital Equipment Corporation (DEC) are bound to contrast their global network of service offices and well-trained network engineers with FORE's small cadre of support professionals. For customers with branch offices all over the world who need ATM network integration and support for their mainframe and minicomputers in enterprise networks, IBM and DEC might offer a more comfortable solution. This is particularly true for customers who already use IBM or DEC for their computer maintenance. The desire for a one-vendor maintenance solution for all enterprise network needs will probably be the toughest argument FORE Systems' salespeople will have to overcome in order to maintain their market dominance.

Summary

FORE Systems had dominated the ATM market from the beginning. It has offered proprietary solutions wherever the ATM Forum has failed to provide industry standards. While its first switches were sold primarily to universities and other research institutions, it now feels that ATM standards and technology have progressed to the point that true commercial applications are possible. It offers both ASX ForeRunner workgroup and backbone switches. The ForeRunner LAN Access switch acts as a router so legacy LANs can be part of an ATM network.

FORE switches are modular in design and distribute the processing power throughout a network. FORE Systems believes that this approach increases system fault tolerance as well as productivity. Workgroup switches can be differentiated from backbone switches because their configurable ports are much smaller in number and because they feature only LAN interfaces and not wide area network interfaces.

All switches come with the ForeThought internetworking software. This software includes the latest ATM-Forum-approved LAN emulation. FORE has supplemented the ATM Forum's specifications with many proprietary features. Buffer control and traffic congestion handling are just two examples of areas where FORE has stepped in to offer superior performance by using proprietary algorithms. ForeView network management software is SNMP-based and runs on a number of platforms, including HP OpenView, SunNet Manager, and NetView/6000.

FORE is not a player in the wide area network switch market, but it has a strategic partnership with Northern Telecom so there will be complete interoperability between Northern Telecom switches and FORE System switches. FORE does offer a very wide range of wide area network interfaces for its backbone switches.

6

ATM's Role in an IBM Enterprise Network

In this chapter, we'll take a close look at IBM's vision of ATM's networking role. The company clearly has the most comprehensive range of products, ranging from LAN bridges and legacy switches to wide area network switches. While there are still many holes in IBM's stated plans for ATM, examining these plans reveals how an IBM-centric ATM network compares with one from other vendors.

The Importance of ATM to IBM

How important is ATM to IBM? Although it was a late arrival to the LAN marketplace, IBM has "bet the farm" on its success as a dominant ATM player. CEO Lewis Gerstner has been quoted as saying that ATM will prove to be as significant to IBM during the next 15 years as the 370 mainframes have been in the past. IBM's strategy contains three key initiatives:

- The company will use its breadth of technology to provide end-to-end system solutions from the desktop to the campus to the wide area network. It also plans to work with developers to ensure an ample supply of ATM applications. Network management is a key part of the system solutions.

- The company will deliver a smooth migration for customers from existing technologies, networks, and applications. ATM products will coexist and interoperate with current network products.

- The company supports industry-wide cooperation and will continue to be active in the ATM Forum as well as in other standards bodies.

IBM's view is that ATM will develop in local area networks in general and at the campus level specifically, and that it will be interconnected within the LAN environment.

The company realizes that it will have to support customers' existing wiring, including unshielded and shielded twisted-pair wire as well as fiber. Its 25-Mbps ATM technology runs over voice-grade (category 3), unshielded, twisted-pair wire using the technology from 16-Mbps Token Ring networks. IBM is positioning its 25-Mbps ATM as a low-cost, entry product for early adopters. The company will use partial-response IV, or PRIV, technology developed by IBM Research in Zurich to implement higher speeds across UTP and STP media, which will be scalable up to 155 Mbps.

Switched Virtual Networking

IBM has come to realize that an ATM strategy must include interoperability with all legacy LAN equipment. The result is what the company called *switched virtual networking (SVN)*. It represents a comprehensive model for building and managing switch-based multiprotocol networks, including, of course, ATMs. The plan calls for building multiprotocol routing support as well as management support into IBM's switches. This protocol support is known as *multiprotocol access services (MAS)*, and it hasn't been created yet. By distributing the routing function so there are MAS servers as well as MAS clients scattered throughout a network, IBM hopes to eliminate a single point of failure.

What's also clear is that IBM is planning to leverage its hardware technology and try to convince customers that a switch-oriented, one-vendor solution is far superior to what all other vendors can offer. It also intends to open up its management support to include support for other major network management packages, such as OpenView.

IBM intends to implement switched virtual networks in three phases. The first stage consists of multiprotocol/hub switch integration with the 8260 hub. The second stage consists of integrating SVN's multiprotocol support with the company's LAN switches. The third stage is distributed route switching.

Figure 6.1 shows IBM's vision of switched virtual networking. Network management and the ATM support services known as broadband network services (BBNS), an architecture I'll discuss next, both run through the entire network. Notice that at the periphery are mainframe servers, ATM workstations running 25-Mbps ATM, and legacy LANs with Ethernet and Token Ring switches.

Right now there are no standards for virtual LAN interoperability. IBM has promised that SVN will support configuring virtual networks by ports, MAC addresses, and by protocol. Quality of service will be supported on these virtual LANs.

IBM argues that switched virtual networking is scalable from 25-Mbps ATM desktop applications to campus backbone and wide area network solutions that will exceed 600 Mbps. Equally important is IBM's decision to speed up performance by placing routing and many other network services in silicon so they're hardware-centric rather than software-centric. Finally, IBM's endorsement of Token Ring and Ethernet switching and its decision to provide these switches with full ATM interoperability is crucial to its large customers. Many of these customers were afraid that their major commitment to Token Ring has saddled them with dead-end technology. Now the company is promising them a smooth migration path that leverages the existing technology by beginning with Token Ring switching, includes 25-Mbps ATM for multimedia and video

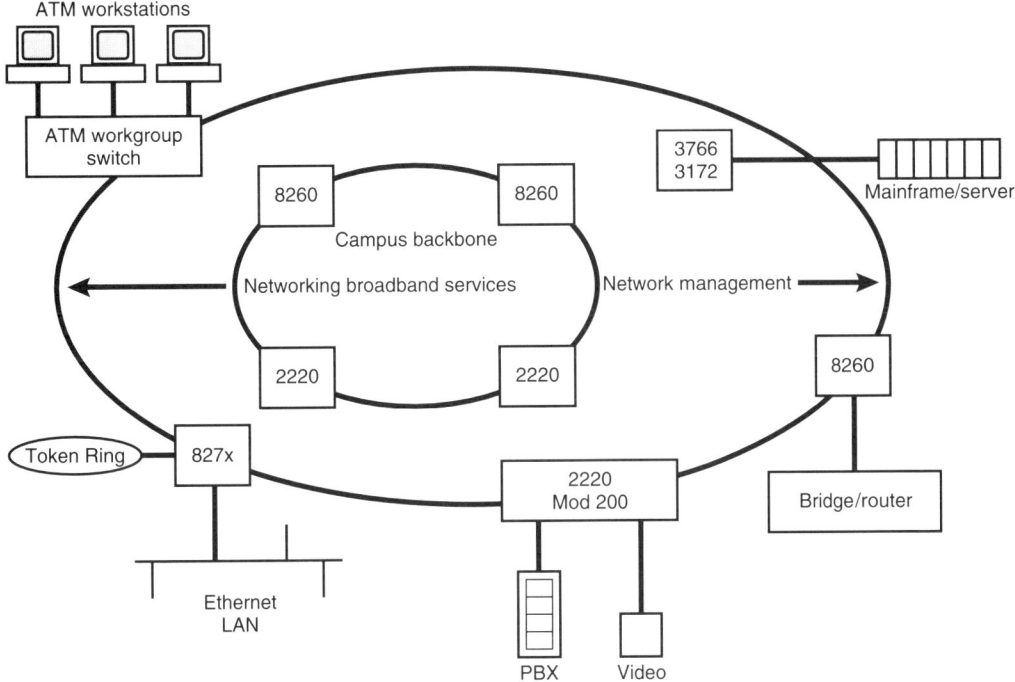

Figure 6.1 IBM's vision of switched virtual networking.

applications, and then moves to ATM switches that are fully interoperable with IBM's campus network hub, the 8260.

Broadband Network Services (BBNS)

IBM started with its vision of ATM campus and wide area network services and then filled in the gaps with switched virtual networking. Its ATM architecture plan is known as *broadband network services (BBNS)*. It includes access services, transport services, distributed network control, intelligent directory services, dynamic bandwidth allocation, guaranteed quality of service, nondisruptive rerouting of connections around network link failures, and congestion control.

IBM's comprehensive ATM strategy includes both 25-Mbps and 100-Mbps adapters, ATM workgroup concentrators, ATM switches, and switch network control management that runs on a NetView/6000. IBM's ATM plan includes expanding its 8260 intelligent hub to include ATM switching capability, developing integrated bridging and LAN emulation servers, and developing a 100-Mbps ATM adapter for its 3172 mainframe controller, as well as an interface for its intelligent hub.

IBM's "Switch on a Chip"

At the heart of all these products is IBM's "switch on a chip" design, developed at the IBM Research Laboratory in Zurich. This chip features 16 input ports and 16 output ports, all of which operate simultaneously. There's a high port count on a single chip, so fewer chips are needed for larger switches. As an example, a 32-port single-stage switch requires only four switch-on-a-chip chips rather than the 64 chips it would need if they had only four input and four output ports each. Each port is capable of a 300- to 400-Mbps transmission. This single chip can drive more than 6.4 Gbps of aggregate throughput and work in conjunction with multiple chips to produce higher throughput. The future development of this chip, as laid out by IBM, promises a throughput of 50 Gbps. IBM's chip is scalable, depending on the number of multiple switch elements that are linked. When linked, these chips share memory and built-in flow control.

The switches are nonblocking and contain built-in support for *automatic load-sharing*. A 622-Mbps ATM link can carry traffic from more than four lower-speed ATM access links. Hardware manages the bandwidth on links. Four switch-on-a-chip ports can be combined to support a quadruple-speed link without the need for any software to control which connection is allocated to a specific switch on a chip because it's handled by hardware.

Another major feature for the switch on a chip is its support of multipoint connections for such services as video distribution and teleconferencing. Copies of a packet can be broadcast to all a switch module's output ports.

IBM believes that its switch on a chip is an improvement over conventional ATM chip designs because it separates data and control information and thus optimizes the function of each section. Figure 6.2 illustrates the chip's structure. The switch on a chip is IBM's answer to FORE Systems' smart buffers. The chip dynamically shares its buffer space among all outputs, while maintaining logically separate output queues. Built-in flow control handles those situations when traffic exceeds the buffer's capacity. Whenever the size of the output queue exceeds the threshold or

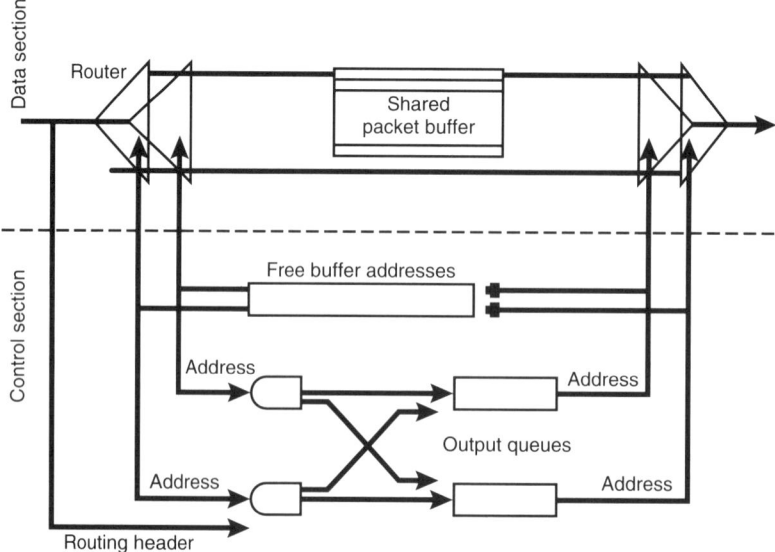

Figure 6.2 The structure of IBM's switch on a chip.

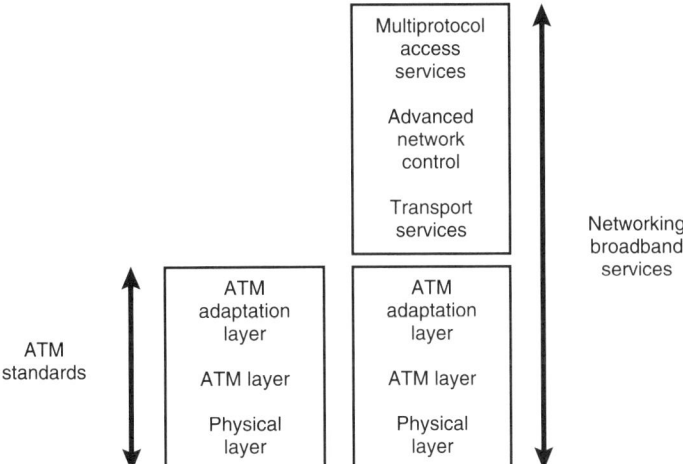

Figure 6.3 BBNS architecture.

when the chip's global shared packet buffer becomes full, a flow control signal is is-sued that indicates the packet isn't accepted. So when the flow control signal is in place, the switch on a chip operates without packet loss.

Networking Broadband Services (NBBS) Architecture

As mentioned earlier, IBM's architecture for high-speed networking is known as *networking broadband services (NBBS)*. The key concepts associated with this architecture include distributed network control to eliminate bottlenecks and in-telligent directory services to reduce network control overhead in locating users. Bandwidth is allocated dynamically, permitting the handling of isochronous traffic and multicasts. NBBS provides guaranteed quality-of-service measures as well as nondisruptive rerouting of connections around network-link failures.

IBM stresses that, while the network management and control features add value to a high-speed IBM network environment, the ATM support is all standards-based and not proprietary. Think of the NBBS architecture as a series of layered protocols that include access services, transport services, and network control services. Figure 6.3 reveals the NBBS architecture and shows how it supplements ATM standards where specifications are not yet in place.

NBBS access services

NBBS access services support the emerging protocols and services offered by an NBBS network. IBM has developed access agents that provide access services by translating the external supported protocols, such as HDLC and IP. External nodes connected to NBBS can communicate with its access services via native protocols without knowledge of NBBS features. Different types of nodes can communicate, including APPN, IP, frame relay, and direct ATM connection nodes. These access services are smart enough to understand and interpret the external services or protocols. They can also locate the target resources by performing address resolution. They maintain and take down connections across a network in response to connection requests, as well as manage bandwidth to ensure fairness among users.

NBBS transport services

NBBS transport services provide transport across the network for user traffic generated at the edges. They perform transmission scheduling and hardware-based switching with multicast capability. NBBS transport services support ATM switching as well as automatic network routing, and match appropriate services to QOS classes. Both ATM point-to-point and point-to-multipoint connections are supported. Transport services can handle constant bit rate, variable bit rate (real time and nonreal time), and best-effort traffic simultaneously. Traffic is assigned to one of four buffers, depending on its type. This prevents high-priority traffic from being forced to wait behind lower-priority traffic. Until all traffic is native ATM cells, IBM has to deal with the problem of transmitting variable-sized packets as well as cells. Its solution is an algorithm known as *preemptive resume mechanism*, which ensures that short packets and ATM cells move ahead of longer packets. After they've been transmitted, the longer packets are allowed to resume their path.

Customers of the 2220 Nways broadband switch can choose whether they want to operate the switch to support both ATM cells and variable-length packets or whether they prefer to support only ATM cells. The first option permits both frame relay and IP traffic. For this type of traffic, IBM supports *automatic network routing (ANR)*. This is a source-routing algorithm that moves control information through the network smoothly by concatenating labels found at the front of each packet. Each label contains information needed to move the packet along to the next node. As the packet moves through the network, these labels are stripped off and the next label is revealed. For acknowledgments, IBM provides ANR with *reverse path accumulation*. This means that, as the packet moves through a network, it accumulates information. The acknowledgment goes in the opposite direction and strips off a label as it reaches each node along the way. The option of transmitting only ATM cells over a NBBS network is probably a few years off for most users.

NBBS control services

NBBS control services control, allocate, and manage network resources, including bandwidth reservation, topology updates, and group management support, for multipoint connections. This set of services identifies new resources and users automatically, supports multiple virtual private networks and multiple virtual LANs on the network, and assists in establishing and maintaining multipoint connections.

Network control services provide directory service support as well as multicast group management support. This layer is also responsible for setting up and taking down connections as well as monitoring and rerouting connections. For routing and rerouting connections, NBBS control services use a topology database to find each node's location in the network as well as what links exist and how much traffic is already loaded on the links using a path selection algorithm. At the same time, it also takes into account the requested quality of service to determine the correct amount of bandwidth and the minimum number of hops required for data to reach its destination.

Congestion control is provided by using the "leaky bucket" approach approved by the ATM Forum and mentioned previously in chapter 5. Using this approach, packets that enter the network in bursts are provided with tokens. When the tokens run out, packets without tokens are designated as discardable if there's a congestion problem. The "leaky bucket" term refers to the similarity between the packets being discarded and a bucket leaking water when it's overfilled.

IBM's NBBS adds value to congestion control by adding a filter to the leaky bucket approach. The filter can analyze buffers and increase or decrease the amount of bandwidth associated with each connection. This adjustment takes place in real time.

NBBS provides additional bandwidth management capabilities. It computes what IBM calls *equivalent capacity*. This value is the fraction of a network link's capacity that's set aside to support a particular connection and meet its quality-of-service requirements, such as end-to-end delay, throughput, and cell-loss ratio. IBM's analysis of video traffic over an NBBS ATM network using bandwidth management shows a savings of up to 50% bandwidth on a single connection. It argues that this savings in bandwidth can be translated directly into dollar savings.

To distribute information such as bandwidth reservations, NBBS uses a technique called a *control point (CP) spanning tree*. The CP algorithm is distributed in each node, and each node that's brought up on an NBBS ATM network views itself as the leader of its own CP spanning tree. Other links joining the network combine and also determine which of these nodes is the leader. Eventually the tree expands until it encompasses the entire network. If a link goes down, the CP spanning tree rebuilds itself using whatever links still exist. The reason for CP is that IBM believes it distributes network topology and other information much more quickly than broadcasts messages because only the header information needed to transmit it to the next link is examined. Topology is updated continuously as control information flows through the network. IBM also believes that CP spanning trees require transmission of less control information because only a very small subset of the total nodes on a network need to exchange the information. The combination of CP spanning tree and NBBS's bandwidth management algorithms means that it's possible to have nondisruptive path switching that's transparent to the users of a network. It also means that connections can be maintained even when there are link failures.

The Nways Broadband Switch Family

IBM renamed its transport network node (TNN) the 2220 Nways switch. Customers can deploy this family of switches in private networks and need purchase only bandwidth, as opposed to ATM services from carriers. IBM is also trying to convince carriers that the 2220 Nways switch should be the basis of their ATM offerings. IBM plans a local area network ATM switch. Wide area models include the 200, a 200-Mbps switch with T3/E3 interfaces, the 300, a six-slot, 2.1-Gbps switch with OC-3 ports, and the 500, an eight-slot, 4.2-Gbps product with OC-3 ports. The 501 is a six-slot expansion unit for the 500. The model 700, a 40-slot, 25.6-

Gbps switch, and the model 800, an 80-port, 51.6-Gbps switch, are both geared for public service providers.

Each Nways switch slot has about a 270-Mbps switching capacity and can accommodate several different types of adapters ranging from low speed (up to 2 Mbps according to IBM) to high speed (6.3 to 52 Mbps). The 500/501 product is built around a 16×16 cell switch. Each adapter can run full duplex at up to 266 Mbps. Simple mathematics indicates that the eight-adapter model 500 will have an aggregate throughput of 2.1 Gbps with an I/O rate of 4.2 Gbps, while a 16-adapter version will have throughput of 3.7 Gbps and an I/O of 7.4 Gbps.

The IBM 2220 Nways broadband switch provides 4.2-Gbps switch fabric throughput. It contains the following port interfaces:

- V.35, V36 low speed
- V24, X21 low speed
- T3/E3/E2/J2
- ATM/UNI T3/E3
- ATM/UNI OC3/STM1
- Multicast ATM
- ATM/UNI 100-Mbps TAXI
- PVC and SVC connections

Trunk interfaces available with a voice server adapter include the following:

- V35, X21, fractional T1/E1/J1
- ATM E3/T3
- ATM OC3/STM1

The Nways switch products are members of IBM's family of multiservice transport nodes designed for wide area networks. The switches in this family are nonblocking and self-routing. It's possible to install two switches in a node to achieve fault tolerance. To maintain a high-availability network, the switches support hot plugging and unplugging of cards, machine upgrades, and maintenance without disrupting operations. Network management is provided by software running on the NetView/6000 platform. Because the software supports simple network management protocol (SNMP), it can be used to manage other products as well as the ATM switches.

Nways switches offer voice compression, silence removal, fax detection, and digital echo cancellation on 64-Kbps pulse-code modulation voice channels. With the voice extension modules, switches can support from eight to 140 voice channels. One argument that IBM will use when positioning these products for private network customers is that its efficient handling of voice as well as data means that voice traffic savings will help justify these large switches. One of IBM's major selling arguments is that, particularly for private network wide area users, its effective handling of quality of service and bandwidth could reduce recurring bandwidth costs for typical customers by up to 50%.

IBM's switches support frame relay as well as ATM, so customers can begin with frame relay and then migrate to ATM later if they want. IBM has positioned its high-end switches as ideal for carriers as well as large customers. The company will probably also add features designed to attract carriers, such as integration with cable TV services, directory lookup, and customized billing and accounting.

ATM Adapters

While most IBM product families associated with ATM share the Nways label, ATM adapters have the name TurboWays. The TurboWays 25 supports the 25-Mbps desktop ATM technology for which IBM fought so very hard within the ATM Forum to win approval as a standard. The TurboWays 100 ATM adapter supports NetWare 3.12 and higher versions, and also supports Novell's ODI interface for network interface cards. There's a TurboWays 100 adapter designed for RS/6000 systems running AIX/6000 V3.25 and higher. Not surprisingly, IBM has TurboWays 25 and 100 cards to support OS/2 and a LAN server with NDIS 2.01 and higher. IBM's ATM adapters, regardless of their speed, will support only LAN emulation and not native ATM applications until at least the third quarter of 1996.

The Role of 25-Mbps ATM

IBM has been the leading advocate of 25-Mbps ATM. Figure 6.4 shows a typical desktop workgroup using 25-Mbps ATM. IBM makes several arguments in favor of this technology. First, it matches the capabilities of many of today's PCs that have only ISA buses. PCs without higher-speed buses can't use 100-Mbps ATM anyway. A second argument is that 25-Mbps is based on proven technology, in this case Token Ring; it's very hard to argue that point since Token Ring technology is known for its reliability. Because the technology supports full-duplex, dedicated links, there's the possibility of 50-Mbps bandwidth and, unlike shared bandwidth such as Ethernet, consistent response time.

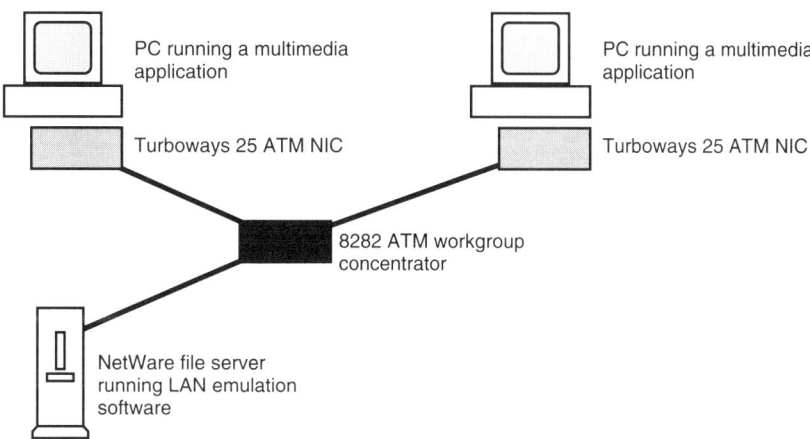

Figure 6.4 25-Mbps desktop ATM in action.

Equally important is that many major customers of Token Ring LANs feel they now have a dead-end technology. Because ATM 25 is based on Token Ring technology, it provides a migration path toward higher-speed ATM. The ISA-based TurboWays 25 ATM adapter must be used with an IBM 8282 ATM workgroup concentrator or a switch that supports 25-Mbps ports. A minimum of a 80386-based PC and a 16-bit ISA bus is also required. This adapter comes with software that supports diagnostics, virtual connection setup and teardown, and bandwidth allocation and management. It also supports ODI and Token Ring LAN emulation. For this emulation, both the TurboWays 25 and the 100-Mbps version (the TurboWays 100) require a PS/2 NetWare server with the LAN emulation function operating and supporting these adapters. A maximum of one card can be installed in a system.

The TurboWays 100

The TurboWays 100 is designed primarily for servers and was developed to transmit voice, video, and data. This adapter is based on the Micro Channel Architecture (MCA) and supports servers running NetWare 3.12 or higher and 4.01 or higher, as well as ODI drivers. It's also designed for RS 6000 workstations. LAN emulation is provided for Token Ring LANs. PCs require a 32-bit, type-5 MCA bus slot to use this card. All the features described for the TurboWays 25 adapter card also apply to the TurboWays 100 card. A maximum of two cards can be installed in a system. This card supports ATM adaptation layer 5 (AAL-5) and configuration with the graphical user interface-based RS 6000 system management interface tool, and supports both PVC connections and ATM-Forum-compliant SVC signaling. There's also support for TCP/IP and simple network management protocol (SNMP) through an SNMP agent.

Integrating ATM into the IBM Mainframe Environment

One key stumbling block in integrating ATM networks with IBM mainframes is bandwidth mismatch. The fastest channel speed supported by IBM is 136 Mbps. This means that 100 Mbps and not 155 Mbps will be the maximum ATM speed supported for quite a while. Even support for 100 Mbps will require significant amounts of buffering as well as some technical tricks to "throttle" back data speed.

ATM and APPN

IBM has developed specifications for migrating existing advanced peer-to-peer networking (APPN) applications to ATM. These specifications map IBM's high-performance routing (HPR) directly to ATM's quality-of-service specifications. These specifications will probably be implemented in routers, hubs, and other devices that support HPR sometime in 1996. Such an implementation would enable APPN/HPR users to use APPN's class of service regarding route security, transmission priority, and bandwidth between session partners across an ATM network. Users could also migrate to ATM services by installing ATM adapters on their ATM network access devices.

APPN/ATM internetworking is particularly appealing because it allows users to deploy the same SNA class of service routines across an ATM network without having to change their applications. The key to this interoperability is the extension of HPR so it can use and request the different kinds of ATM services.

Linking LANs and Mainframes to ATM Backbones

IBM's 8281 ATM LAN bridge connects the ATM and LAN worlds. For hosts, a 100-Mbps ATM adapter can be placed in an IBM 3172 interconnect controller. It's also possible to have direct attachment to the IBM scalable POWERparallel system with a 100-Mbps ATM adapter.

Figure 6.5 reveals IBM's LAN emulation plan for its customers. It permits Ethernet and Token Ring data to flow across an ATM network. The network management is provided by NetView for AIX. Simple network management protocol (SNMP) is the vehicle for providing key management information. IBM's SNMP agents are found on its adapters, hubs, bridges, and concentrators. The NetView for AIX is attached to an 8281 ATM LAN bridge with transparent source-routing bridge capability. The 8281 enables the directory function to perform ATM UNI management.

The TurboWays 8282 ATM Workgroup Concentrator

IBM's 8282 ATM workgroup concentrator is part of the company's 25-Mbps desktop ATM solution. It can connect up to twelve 25-Mbps ATM workstations in a single ATM

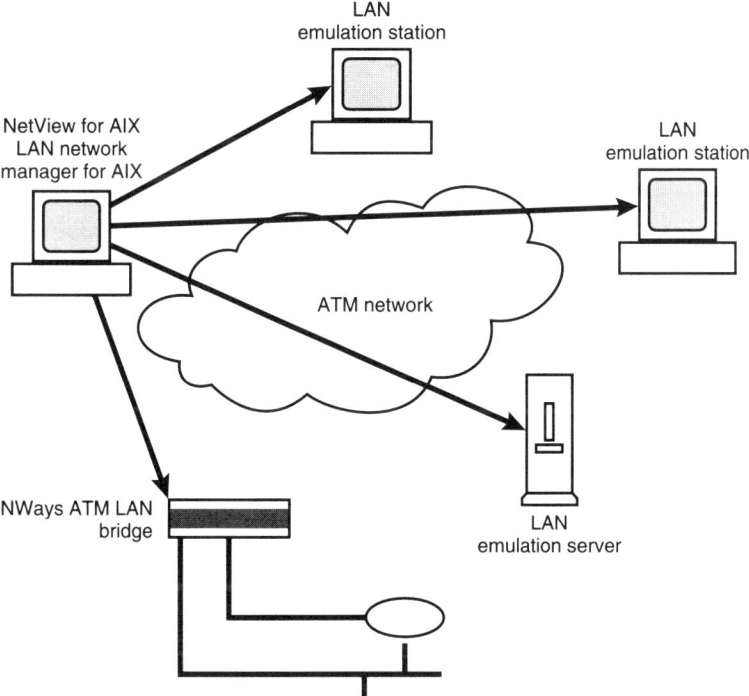

Figure 6.5 IBM's vision of LAN emulation.

switch port. There are 8-port and 12-port versions available. A 100-Mbps ATM switch port is also included, as well as an RS-232 serial port. The concentrator can be managed via SNMP and supports both unshielded and shielded twisted-pair wire. While IBM's initial 25-Mbps ATM offerings support only LAN emulation, the company has indicated that it will support either PVC or ATM-Forum-compliant SVC Q.2931 signaling.

The Hub as the Focal Point of the Network

IBM wants to own the campus backbone, which it sees as the heart of the enterprise network. Its hopes rest on its 8260 multiprotocol intelligent switching hub. The hub is manufactured by Chipcom/3Com, but the ATM technology in the form of ATM modules is all IBM. The company has stated that this hub is "the platform on which ATM technology will be integrated to meet the network needs for the next 5 to 10 years." The hub features IBM's switch on a chip, discussed earlier in this chapter. IBM argues that the hub is designed to handle all enterprise networking needs because it supports direct attachment of servers, high-performance workstations, other hubs, and LAN bridges. It also supports subnetworking (see chapter 3) and 25-Mbps desktop ATM.

An SNMP-based control program that comes with the 8260 hub is responsible for network topology. It computes the routes cells will take and builds the VPI/VCI tables that control ATM switching. Figure 6.6 shows the role IBM's ATM switching hubs play in its vision of the enterprise network in general and the campus backbone specifically.

The 17-slot 8260 now comes with ATM, but you can upgrade a conventional model to ATM functionality by adding an integrated ATM backplane. The ATM backplane supplements the existing backplane so ATM modules can be interconnected on it while the conventional backplane can interconnect other types of networks. Additional ATM support includes a combination ATM switch/control point module and a 100-Mbps ATM fiber concentration module. The 8-Gbps capacity switch supports redundant modules, fault tolerance, hot-pluggable modules, and per-port switching.

The 8260's new backplane contains separate connectors to an 8250 compatible bus as well as to an 8260 extended bus. The link to the older 8250 bus means that customers with 8250 hubs will have full compatibility with the three Ethernet seg-

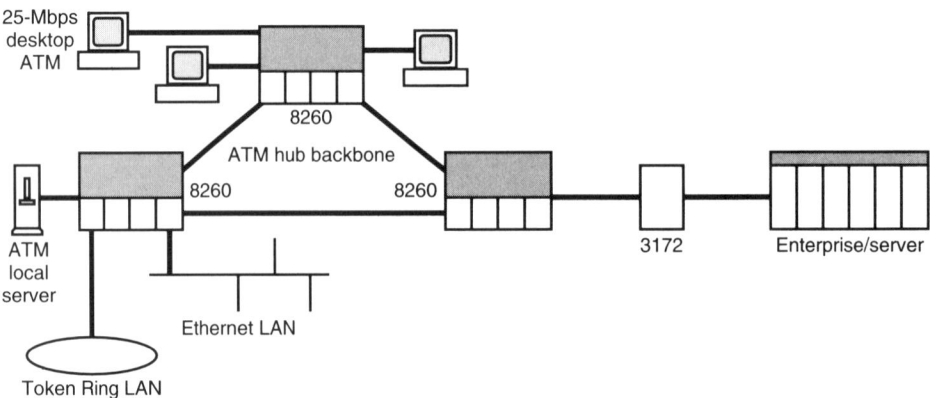

Figure 6.6 IBM's switching hubs and the campus backbone.

ments that can be linked to the 8260. The 8260 extended bus offers a set of point-to-point links between adjacent slots, optimized for creating backplane rings. This new bus is capable of supporting 2 Ethernet, 10 Token Ring, or 4 FDDI networks, available to 8260 modules. For LAN-centric environments, the 8260 has an extended backplane capability of up to 8 Ethernet, 17 Token Ring, and 8 FDDI networks.

The switch/control point module offers switching functions as well as network control functions, such as setup processing, route selection, and topology discovery, in one module. The circuit card for this module is a single-chip integrated circuit configured as a 16 × 16 times 8-bit parallel switch with 8 Gbps of aggregate throughput. Both in-band and out-of-band management is supported. The module self-learns by discovering network resources, topology, etc., and then bypasses failed nodes and links. The module comes with troubleshooting tools and incorporates flash memory so code can be updated in the future. Both permanent virtual-circuit signaling and switched virtual-circuit signaling are supported in both point-to-point and point-to-multipoint nodes. The module is designed to facilitate connecting local ATM networks into a WAN and comes with a permanent virtual path with throughput control to help build this network. This module requires two slots.

The 100-Mbps fiber concentration module is designed for campus backbone communications between two hubs as well as for parallel links to increase bandwidth. Eventually, when software is available, it will also serve as a high-capacity link for servers in native ATM mode. Application-specific integrated circuits (ASICs) supply the traffic management and congestion-avoidance functions.

The 8260 offers distributed management architecture to reduce hub cost and enable network managers to configure multiple segments. DMM provides the same network management functions performed within the 8250, plus some new ones. The network management functions carried over from the 8250 include configuration management, power management, SNMP agents for the entire hub, local terminal management (out-band), remote management via Telnet (in-band), and communication with all media access cards for network statistics and interpretation. IBM has packaged what it calls the *distributed management function (DMM)* on different modules. A stand-alone module (DMM) uses a media access card to monitor LANs and gather statistics and must be plugged into an 8260 media module. A DMM with an Ethernet carrier (EC-DMM) starts with the basic DMM functions and then adds an interface to Ethernet backplane networks. It can interface with up to six Ethernet media access cards and monitor these networks concurrently.

Network Management

While IBM offers several different network management packages, they're all part of an overall network management plan for ATM networks. IBM has committed itself to meeting certain network management needs; it believes its ATM customers want the following features, which can be used in conjunction with each other:

- Topology management to determine the physical and logical network structure
- Change management to alter network configuration dynamically or on a scheduled basis

- Fault management to identify network problems

- Operations management for performing remote network operations

- Accounting management to keep count of key network elements such as the number of ATM cells using a particular circuit

- Performance management to gather key information on how the network is performing, as well as to model future behavior

- Standards-based solutions to ensure that products purchased won't become obsolete or lack interoperability with other network products

Nways Broadband Switch Management

The Nways switch control program runs on each ATM node and includes the code required to run the Nways switch. This program implements the NBBS architecture to support access services and protocols. The program can be accessed from the node itself or from a remote maintenance console.

The Nways Broadband Switch Manager for AIX

This software runs on NetView for AIX platforms to gather and consolidate management information from broadband switches. It provides notification of any additions, removals, or change in the status of nodes, trunks, and adapters. The software displays an icon-based network topology map that's color-coded. When an object is selected, a window opens and displays information about that object. The network manager can manage switches on an ATM network using this software and respond to problems by issuing commands.

Enterprise-Wide ATM Network Management

IBM touts its NetView/390 platform as ideal for a network-centric organization, and has positioned NetView for AIX to be used for distributed networks. According to IBM, the NetView/390 platform is designed for managing very large networks with heterogeneous transport and management protocols. NetView/390 and NetView for AIX communicate with each other over an LU6.2 SNA/MS session. The NetView/390 platform is restricted to performing only topology, fault, and accounting management. Only the distributed manager, NetView for AIX 1, communicates with the larger machine to prevent smaller workstations from flooding it with information. Figure 6.7 shows IBM's reference configuration to explain the path its network management information takes.

Nways Intelligent Hub Management Program for DOS (IHMP/DOS)

This SNMP-based program offers a graphical user interface for managing small- to medium-size networks that use IBM hubs. It runs on a PS/2 compatible computer and provides the network manager with the ability to perform the following tasks:

- View or change hub configurations
- Gather and display network statistics at network and port level
- See the network in a color-coded display
- Manage the hubs remotely through the program's support for RMON

Nways Manager for Windows

This package combines and integrates management for IBM's hubs, switches, bridges, routers, ATM concentrators, and Ethernet- and Token-Ring-to-ATM bridges. The software uses a graphical user interface and comes with a NetView for Windows license. The program is designed specifically for small- to medium-size networks (25 to 250 nodes) as well as for remote offices. IBM stresses that the software is self-contained and doesn't require purchasing any additional products. The software allows you to perform the following tasks:

- View and change subsystem configurations
- See problems in color for easy identification
- Set thresholds for error notification
- View a graphical display of the entire network
- Monitor real-time events and display a time-stamped alarm log
- Gather information with integrated trouble ticketing in order to track problems from detection to resolution
- Use a MIB browser for management of components not supported with a graphical interface

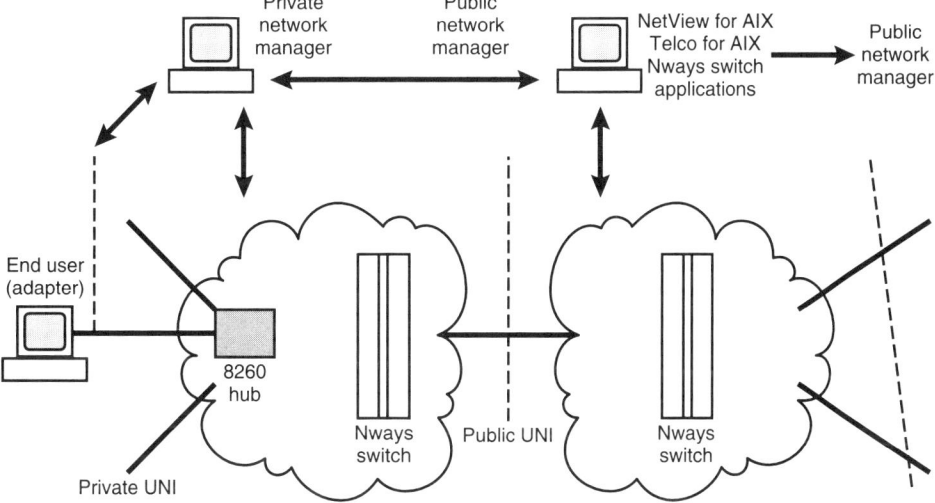

Figure 6.7 The path network management takes in an IBM ATM network.

IBM's Future ATM Network Management

While IBM is short on details, it has indicated that it will consolidate many of its network management packages to make them less confusing. It also plans to open up network management by including agents that can be accessed by other vendors' major network management packages. Finally, it plans to license its management technology to other vendors.

Is IBM the Right ATM Vendor for You?

IBM brings enormous resources to the ATM marketplace. It clearly offers the most complete one-vendor solution for ATM users. By leveraging its strength in the areas of network management and chip technology, the company is able to clearly position itself in the ATM marketplace. IBM insists that all its networking products, including hubs, switches, servers, and mainframes, can be managed via NetView and SystemView management tools. All its products can also be managed via the simple network management protocol (SNMP) and the common management information protocol (CMIP).

Furthermore, if a company is even contemplating LAN-to-WAN integration with ATM, then IBM offers its private wide area network ATM switches and a promise of complete interoperability from LAN to WAN. Because these switches also support frame relay, a company can begin with frame relay and then migrate to ATM sometime in the future.

Are there any negatives associated with choosing IBM as your ATM vendor of choice? Perhaps the major one at this time is that so much of IBM's overall vision for ATM networking is still on paper; the missing links, especially switched virtual networking, are significant. Also, IBM has not been a player in the workgroup switching market, choosing to begin with the wide area network category switches and later develop smaller switches. So a company that isn't looking at enterprise networking in a campus environment but prefers to keep ATM switching at the workgroup level might be more attracted to a vendor that has been emphasizing this market, such as FORE Systems. The other type of customer with some reservations about ATM are probably companies who have invested heavily in hub and router solutions from other vendors. Rather than having IBM 8260 campus hubs, they might have Cabletron or Bay Network equipment. These companies, because of the money invested in these products, might want to look to their major vendors for ATM support.

Summary

ATM is absolutely crucial to IBM's future success as a company, and it has committed its vast resources to this area. The company believes it has the most comprehensive, scalable ATM solution. It also stands alone in offering a smooth migration path to companies with legacy LANs by enabling them to begin with Token Ring and Ethernet switching, add 25-Mbps ATM for desktop users who require it, and then move to ATM backbones and private ATM wide area networks. The company's vision of switched virtual networking (SVN) incorporates multiprotocol routing access services (MAS) in a client/server manner so routing is distributed to hardware across the network.

IBM's networking broadband services (NBBS) is the company's layered ATM architecture of services that includes access services, transport services, distributed network control, and intelligent directory services.

The foundation of IBM's ATM hardware is its switch on a chip. This chip's density means that less chips are required. It provides nonblocking and load-sharing switches. Data and control functions are separated for optimized performance. IBM's TurboWays ATM adapters currently support 25-Mbps and 100-Mbps speeds. While they currently support only LAN emulation, they plan to support native ATM applications in the near future.

IBM believes that ATM will grow first as a campus backbone, with the 8260 enterprise hub as the heart of this network. This hub will support ATM as well as legacy switching and legacy LANs.

Network management from IBM provides management of topology, changes, operations, faults, accounting, and performance. While the company currently offers several different network packages for platforms that include its mainframes, the RS 6000 minicomputer, and PCs, it's moving away from proprietary NetView solutions. It plans to build network management agents into its hardware that can be accessed by non-IBM equipment running major network management packages such as OpenView.

7

Digital Equipment Corporation and ATM

Digital (also called DEC) has quietly been revamping its entire company. The computer industry media has been filled with the financial reverses the company suffered while it shifted from traditional minicomputers to its new Alpha-based machines. What has been largely ignored, though, was how thoroughly Digital developed a comprehensive network product line that includes desktop, departmental, and enterprise switches, support for both legacy LAN and ATM networks, and integrated network management. In this chapter we'll take a close look at Digital strategy as well as its actual products. It has perhaps the broadest offering of virtual LAN switches; hubs that stack, rack, and fit into a chassis; and routers that support virtually all major protocols. Let's see how Digital sees the current network environment and how it proposes that its customers prepare for the next round of technology changes.

Digital's Strategy

Digital is committed to supporting multiple legacy LAN protocols (Ethernet, Token Ring, and FDDI) and developing a seamless ATM enterprise network that spans both the LAN and WAN worlds—one that includes voice, video, data, and imaging applications. The company's strategy initially targets workgroups running data-intensive applications. The second phase adds support for legacy LANs and PCI-bus servers into an ATM LAN backbone. The final stage brings ATM to PC desktops.

The enVISN Architecture

Digital's *enterprise virtual intelligent switched networks (enVISN)* is its comprehensive architecture and plan for the evolution of scalable virtual networks. This plan will eliminate the obstacles to seamless networks that include legacy LANs and state-of-the-art ATM networks. Among the business requirements Digital believes enVISN meets are the following:

- Scalable bandwidth with investment protection
- Flexible network reconfiguration
- Secure access to logical workgroups
- Flexible customization of workgroup environments
- Workgroups sharing high-performance servers
- Mobile workers with same LAN access as local LAN users

The company defines a virtual LAN as a limited broadcast, meaning that all members of a VLAN receive every broadcast packet sent by members of the same VLAN, but not packets sent by members of a different VLAN so bandwidth can be optimized and traffic given firewall protection. Under enVISN, Digital defines three classes of VLANs:

Class 1. A set of ports are grouped into a single broadcast domain. Known as "port switching," this approach can be used to dedicate certain ports to specific functions, such as providing secure access to dial-out facilities. In effect, this type of VLAN groups nodes by physical address.

Class 2. A set of nodes are grouped logically into a single broadcast domain by their MAC layer addresses. One practical application is that clients can be mapped to their servers, independent of their locations.

Class 3. A set of nodes are grouped into a single broadcast domain by their common network layer (subnet) address. Sometimes known as a "virtual subnet," this type of VLAN enables network managers to create virtual networks based on protocols they run, such as IP.

What makes enVISN unique is that Digital allows VLANs to be formed that incorporate any combination of these three classes. The company points out that a network with a combination of class-1 and class-2 VLANs would permit the grouping of physical ports and MAC addresses independent of protocol. Such a VLAN would be effective for handling nonroutable protocols such as LAT and AppleTalk.

Virtual Networks

Digital perceives a major problem with virtual LANs because they don't scale very well once they incorporate a large number of users. It argues that enVISN addresses this issue by creating a new element, the *virtual network (VNET)*. It defines a VNET as a collection of VLANs interconnected with high-performance optimum routing. Communication between VLANs within the same VNET takes place via high-performance multilayer switches with integrated routing or by switches with existing routers.

Digital believes that the traditional routing function should be split into two functions: route determination and frame forwarding. It sees the frame forwarding function as one that should be distributed to switches where it becomes integrated with them, and it sees the route computation function distributed to the switches as well. Notice that there's some similarity between this concept and IBM's switched virtual networks (SVN) architecture. Unlike IBM, though, Digital's enVISN breaks with the concept of route servers because the company believes such an approach causes congestion and presents data synchronization problems. Digital seems much further toward realizing this goal. One major advantage it has over IBM is that it doesn't have the distraction of having to divert resources to develop products that incorporate mainframe computers into its VNETs.

The enVISN architecture offers a number of significant features, including the following:

Access to services. enVISN assigns privileges to a user that create a service policy domain, basically at the workgroup level. Users can communicate with each other if they belong to the same service policy domain.

One-hop switching. Using the intelligence built into a multilayer switch, switching decisions are made locally rather than on central routers or route servers. Multiple hops are eliminated because the switch can create a direct switched connection to the destination.

Network Management under enVISN

Digital argues that there are several advantages to enVISN's network management scheme because it centralizes policy administration while distributing enforcement. Policy-based management enables network managers to establish policies rather than have to configure individual devices. enVISN checks policies before distributing them to devices for enforcement. The VNET manager converts policies into topology rules and then sends them to a VNET configuration server. This server checks the VNET topology and runs consistency checks on the policies. When they're verified, the server distributes the configurations to the individual VNET devices for implementation. Figure 7.1 shows a model of the enVISN VNET manager functions. Note that the functions supported by this manager include policy management membership, security, services, topology mapping, event monitoring, and performance monitoring.

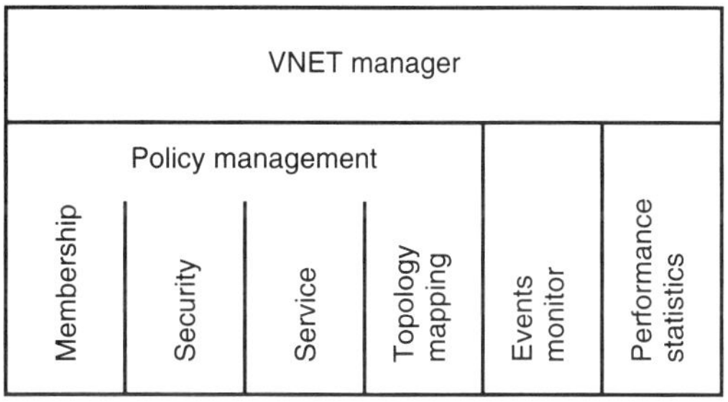

Figure 7.1 A model of enVISN VNET manager functions.

Switches

Digital's switches span desktop, departmental, and enterprise markets. In this section, I'll describe how these switches differ and how Digital has positioned them in the marketplace. The company believes that the desktop cost is crucial. Customers will want to pay a minimal amount of money for data-intensive activities such as database queries and large file transfers to users. The primary reason for desktop switching is to take their power users off contention networks and provide them with their own "express lane" so they don't congest the network and slow down other users.

The switching issues change somewhat at the departmental level, according to Digital. It sees the two major departmental concerns as traffic control and security, the latter via firewalls. From Digital's perspective, the ideal departmental switch must incorporate the security and functionality of a router while still performing at optimum switching speed.

Finally, Digital sees available bandwidth as the primary concern at the enterprise switching level. Enterprise switches provide high-speed links between departmental LANs and their switches. At this level, redundancy is advantageous because it provides some security as well as offering room for growth.

In order for its switches to be competitive, Digital believes they must perform two key functions: configuration switching and LAN switching. Network managers can use software-based configuration switching to optimize bandwidth at the workgroup and desktop levels. Users can be grouped together into virtual networks, described earlier in this chapter, by their current functions and needs rather than by their physical locations. The ability of Digital switches to permit port switching makes it possible to configure these virtual networks.

LAN switching, on the other hand, focuses on the physical connections among LAN users and LAN segments. Digital has chosen to support both store-and-forward as well as cut-through switches because different topologies must be treated differently. Digital believes that store-and-forward switches are better choices for Ethernet because they can detect and reject runt packets and because they carry more robust memory buffers. Because FDDI networks use a noncontention approach, Digital believes that cut-through switching is a better choice for this topology because of the improved network performance.

Digital summarizes its switching strategy by emphasizing its commitment to desktop, departmental, and enterprise network backbone solutions using both store-and-forward and cut-through types of switches. It's also committed to integrating switching, routing, and hub functionality in its network products to provide a smooth migration for legacy LANs to ATM. Finally, Digital's switching strategy identifies the need for a single, easy-to-use graphical SNMP management application on multiple platforms as absolutely crucial.

The DECswitch 400

The DECswitch 400 is a stand-alone, modular, departmental switch that spans both Ethernet and ATM. It supports up to 24 dedicated Ethernet LANs as well as one or two ATM ports. One advantage of this switch is that Ethernet-to-Ethernet traffic doesn't enter the ATM network; therefore, this traffic doesn't require unnecessary segmentation and reassembly. The result is lower latency between connections and no infringement on ATM bandwidth. The switch has an aggregate throughput of 640 Mbps and can support up to 32 VLANs per switch and 65,000 VLANs per network. When used in conjunction with the DEChub 900 MultiSwitch, it will be able to connect enVISN class-1 VLANs in the hub over an ATM backbone. It supports the ATM Forum's LAN emulation client and user network interface (UNI) 3.0 and 3.1 as well as classical IP and inter-VLAN IP routing. The switch can be managed by an SNMP-compliant network management application. Digital offers its own DECswitch 400 network management software, which includes a graphical user interface.

The DECswitch 900

The DECswitch 900 family of store-and-forward switches are designed to provide routing between VLANs in an enVISN environment. They route traffic between the VLANs formed with a PortSwitch 900 module within the DEChub 900 MultiSwitch. Customers choose between IP or multiprotocol packages that provide all routing. These packages are preloaded into flash memory for easy installation. Older DECswitch 900 switches can be upgraded to include routing via simple software upgrade. The multiprotocol package provides concurrent bridging and routing of IP, IPX, AppleTalk, DECnet/OSI, DECnet IV, RIP, and OSPF protocols. The DECswitch 900 family includes switches that support Ethernet, Ethernet and FDDI, and Ethernet and Token Ring. As is the case of Digital's other switches, these switches can stand alone or functions as modules within the DEChub 900 MultiSwitch.

The GigaSwitch/FDDI

This switch provides a 3.6-Gbps switching fabric to support full-duplex, 100-Mbps, FDDI, point-to-point connections in both directions simultaneously. It supports two and four-port FDDI (single-mode fiber, multimode fiber, and unshielded twisted-pair category-5 wire) and two-port ATM linecards. A total of 34 switched LANs can be supported. The switch can be managed using Digital's Polycenter NetView, HUB-watch, or any vendor's SNMP-compliant network management application. Figure 7.2 illustrates how two GigaSwitch/FDDI switches can support a number of different types of devices, as well as communicate with each other via an ATM wide area network service.

The GigaSwitch/ATM

One of the major advantages Digital offers its customers is a clear migration strategy. Its GigaSwitch/FDDI has a migration path to its GigaSwitch/ATM, and the two switches are both compatible with the company's DEChub 900 MultiSwitch. The nonblocking GigaSwitch/ATM networking switch has a 10.4-Gbps switching fabric. Thirteen of the switch's 14 slots can accommodate a four-port ATM module so the switch can support up to 52 155-Mbps SONET/SDH multimode fiber ports. One slot is reserved for a clock/management module. Digital's plans for this switch include support for 155-Mbps single-mode fiber, unshielded twisted-pair category-5 T-3/E-3,

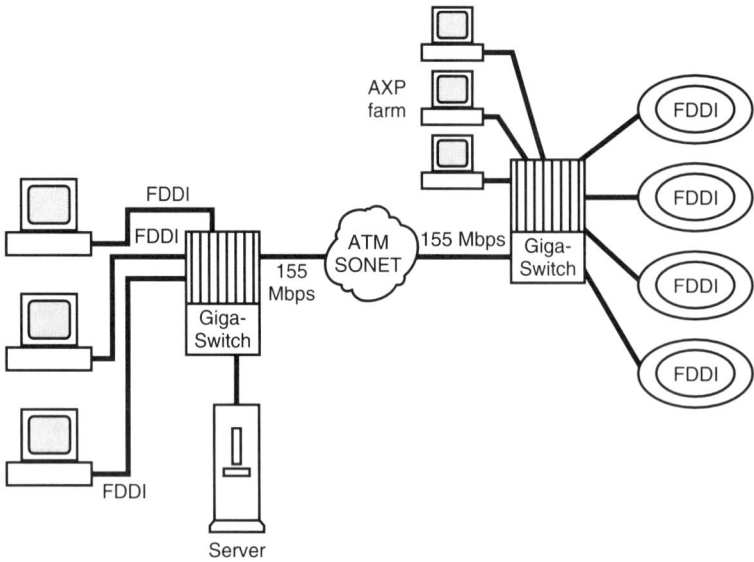

Figure 7.2 GigaSwitch/FDDI switches in an enterprise network environment.

and 622-Mbps SONET/SDH transmission. The GigaSwitch/ATM system supports constant bit rate (CBR), variable bit rate (VBR), and available bit rate (ABR) switching. It can be managed by any SNMP-compliant network management application, including Polycenter NetView and HP OpenView. Among the features offered by the GigaSwitch/ATM system are the following:

Autoconfiguration. The GigaSwitch/ATM is capable of automatically learning a network's topology. If a physical link fails, the switch can reconfigure the network to transmit cells around the point of failure.

Load balancing. The system is capable of choosing a physical route between source and destination that includes the lead loaded link for the new VC.

Dynamic routing. The system uses the shortest-path-first algorithm to route a switched virtual circuit when it's set up and will seek an alternate path if necessary.

SwitchMaster. This advanced queuing management function uses DEC's parallel interactive matching technique to ensure that cells addressed to idle ports are transmitted as quickly as possible without waiting for delivery of all "head of line" cells. The cells stored in input queues and not just the first cell in the queue are matched with an appropriate output port.

Traffic Management

Digital has been part of a group of ATM Forum mavericks who have fought the group's decision to go with a rate-based method (see chapter 2) for traffic management. Digital has joined Ascom-Nexion Timplex, Mitsubishi, and others in arguing for a credit-based approach. Now known as *quantum flow control (QFC)*, this scheme requires reserved buffers. The switch reserves buffers for each virtual circuit (VC) where data is placed, then advertises the current number of unoccupied buffers for a given VC. The problem that many in the ATM Forum find in this approach is that good performance requires a high number of buffers to keep a VC's traffic flowing smoothly. It's much less expensive to use a rate-based approach because it's far cheaper to implement the scheme in silicon than to add expensive memory to switches.

Digital's credit-based approach is known as *FLOWmaster*. Because it hasn't yet been successful in persuading the ATM Forum to accept this scheme as a standard, Digital will offer customers the option of using the standard-based rate approach or its own approach. Digital argues that there are many reasons to choose FLOWmaster.

Switches using the rate-based approach use cell-loss minimization techniques based on a best-effort quality of service. The problem with such an approach is that the switches continue their efforts to retransmit dropped cells on congested networks; Digital believes the ultimate result could be a "throughput collapse."

FLOWmaster, on the other hand, manages traffic flow before a congestion problem can develop. It tracks the number of cells being transmitted so that the sending device maintains a "credit balance" for each virtual circuit. A cell is transmitted only when the sending device recognizes that the receiver's buffer is available to hold it. Digital points out that to use FLOWmaster, both the switch and adapter cards must support this scheme.

Adapter Cards

The ATMworks 750 adapter card is, according to Digital, the fastest ATM adapter in the industry. The company argues that its Alpha-based high-performance workstations are the perfect vehicle for ATM workgroup computing. This card uses 62.5/125 multimode fiber-optic cable and runs at 155 Mbps with SONET/SDH framing. It supports ATM adaptation layer 5 (AAL5), which is designed for class-C connection-oriented traffic. It also supports variable bit rate (ABR) class of service.

The GigaSwitch/ATM System in Action

It might come as blow to some Americans' pride, but the leading country in the world when it comes to implementing ATM is Finland. One of the leaders in the area of ATM research and a member of the ATM Forum is Tampere University in Finland.

Figure 7.3 reveals that the GigaSwitch/ATM is at the heart of the campus's ATM network where it links a number of workstations, servers, and other switches. The network includes several Alpha AXP computers with Digital's ATMworks 750 interfaces, all with FDDI, to create an ATM "supercomputer cluster." The switch also links this cluster to a Cray supercomputer in a Helsinki-area FDDI ring via Telecom Finland's 155-Mbps wide area network ATM service.

Radio and television signals are broadcast over the ATM network as well as over the wide area ATM network. A video-on-demand server runs video conferencing and other multimedia applications and is connected to the GigaSwitch/ATM system. In

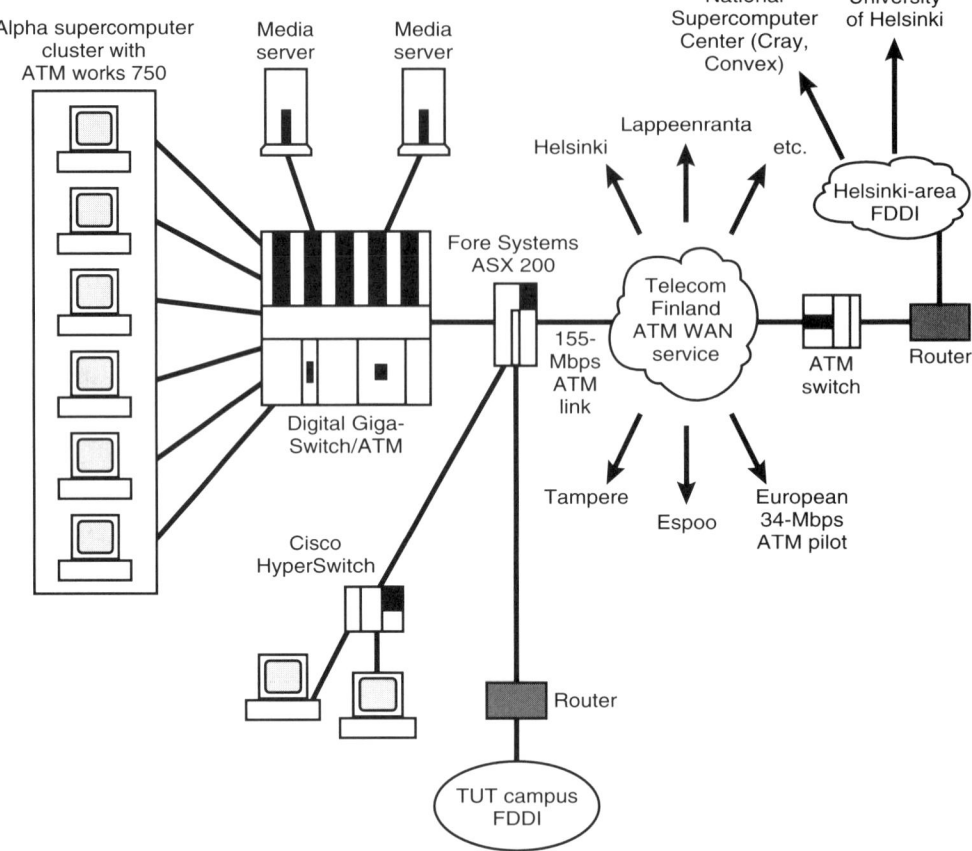

Figure 7.3 A Digital ATM network in Finland. (*Digital Equipment Corporation*)

the past, simulations were generated on a supercomputer and then postprocessed and visualized on a workstation. Now, with the ATM network, the simulations can be visualized in real time. A university spokesperson believes that the best thing about the GigaSwitch/ATM switch is its speed and reports 128-Mbps application throughput with UDP and 133-Mbps throughput with TCP between Alpha workstations using ATMworks 750 adapters.

Toshiba and Digital's ATM Chips

Toshiba and Digital have a partnership agreement to jointly develop highly optimized 155-Mbps ATM segmentation and reassembly (SAR) products for the hub/switch and adapter card markets. Toshiba currently markets Digital's current version of a single-chip 155-Mbps SAR optimized for hub/switch applications. This chip supports the AAL 3/4 and AAL5 interfaces. Digital will use the chip in the future in its DEBhub products.

Hubs

Intelligent hubs play an integral part in Digital's network architecture. What's particularly intriguing about Digital's multistack system architecture is the scalability and flexibility it brings to the DEChub family of products. The hubs are modular, which means they can be installed and managed as stand-alone units, placed as stackable units in a rack-mounted configuration, or arranged in a hub chassis. Companies that start with DEChub 90 hubs can use the same modules in the DEChub 900 Multi-Switch hub chassis.

The DEChub 90

The DEChub 90 chassis functions as a departmental solution. Hubs can be snapped into the chassis by hand without interrupting the operations of other hubs in the chassis. These hubs can be migrated forward to the DEChub 900 MultiSwitch.

The DEChub 900 MultiSwitch system

The DEChub 900 MultiSwitch chassis is Digital's single point of integration for multiple LAN technologies. It has a 3-Gbps backplane that supports 10-Mbps and 100-Mbps Ethernet, FDDI, and ATM. The MultiSwitch supports integrated configuration and dynamic switching at the packet, port, module, and LAN segment levels for building virtual networks. Modules can be hot-swapped as well as load-shared. Up to eight DEChub 90 or 900 modules can be installed in the chassis.

Managing Digital hubs

Digital offers the Windows-specific DECpacketprobe and PROBE watch products for traffic monitoring, fault diagnosis, and network performance tuning on Ethernet and Token Ring LANs. HUBwatch is a graphical management package that enables network managers to configure, monitor, and control a network from a PC or workstation. This software can manage stackable hubs, the chassis-based DEChub 90 and DEChub 900 MultiSwitch hubs, and the GigaSwitch/FDDI and GigaSwitch/ATM backbone switches. HUBwatch can be launched from a number of network management products including OpenView, Novell NMS, SunNet Manager, POLYCENTER/NetView, or Digital's ManageWORKS.

HUBwatch can be used to create and manage virtual LANs within a DEChub 900 MultiSwitch. It can recognize the module and port-level alarms for fault isolation, and traffic monitoring comes from the Digital intelligent agents found within its network products. It has the built-in intelligence to automatically learn secure port addresses and can provide automatic recovery when an FDDI ring is broken.

Routers

The DECNIS multiprotocol router family has a high-speed ATM interface to connect multiprotocol legacy LANs to ATM networks. These routers support several data links including PPP, HDLC, LAPB, Vitalink control protocol (VACP), Cisco HDLC

(CHDLC), and Digital DDCMP. They're manageable from Polycenter NetView and other SNMP management platforms. The DECnis ATM controller 631 is a network interface card for the DECNIS 600 that connects ATM backbones to services such as T 1, frame relay, and switched multimegabit data services.

Should Digital Be Your ATM Vendor?

Digital seems much further along than other vendors in offering a comprehensive product line that incorporates ATM as well as legacy LAN products. Its enVISN architecture provides a virtual LAN bridge between today's legacy LAN world and tomorrow's ATM network. Many factors weigh heavily in Digital's favor:

- Network management with a common user interface integrated into all network products

- Virtual LANs that have the flexibility to offer everything from port switching to MAC address switching based on protocols being run

- Desktop, departmental, and enterprise switches that interoperate

- A one-source vendor for all key network components as well as service

- Value-added technology to supplement gaps in current standards

In contrast to IBM, which also believes in distributing the routing function throughout a network, far more of Digital's products have already been delivered. What is particularly appealing is that the products all seem to have built-in migration paths so customers won't have to throw them away when new products emerge.

Why not Digital? Companies that have a large investment in Token Ring or mainframe hardware might prefer IBM's greater expertise with these products, and companies that already have substantial investments in other vendors' hubs and switches won't be able to benefit from the seamless interoperability that an all-Digital products solution brings. Having said this, the company does support SNMP, and presumably the devices could be managed by a Digital network management program.

Still, Digital has an excellent chance to become the leading ATM company because it has prepared so well by clearly developing a comprehensive strategy. This strategy places an emphasis on flexibility, scalability, and interoperability of all network equipment, including ATM. Give this company a good hard look if you're considering moving to ATM anytime soon.

Summary

Digital Equipment Corporation has developed a comprehensive network product line that incorporates switching at the desktop, department, and enterprise levels and includes virtual LANs, legacy LANs, and ATM networks. The company's enVISN architecture permits port-based switching, segment switching, and MAC address switching. It includes policy-based management so decisions can be made and then implemented throughout a network after being validated. This architecture also includes distributing the routing function throughout a network.

Digital supports Ethernet, Token Ring, FDDI, and ATM switching. Its Giga Switch/FDDI and GigaSwitch/ATM are its enterprise switches. Digital is the only vendor that supports the switching of all these topologies. While Digital has been a firm supporter of a credit approach to ATM traffic congestion management, it will support the ATM Forum's rate-based approach, offering its customers both options.

Digital offers intelligent network management agents in all its networking products so they can be managed via SNMP as well as by other network applications such as OpenView and NetView. The company also offers its own ATM chips, ATM adapter cards, switches, hubs, routers, traffic congestion software, and network management. It has positioned itself as a one-stop shop for all networking equipment including ATM equipment. It currently has the widest range of networking products available.

ATM and
Cisco Systems

Cisco Systems has grown overnight from a router-only company to a company that believes it can offer one-stop network shopping, particularly when it comes to a switched environment. It has filled out its family of products by acquiring a number of key companies such as LightStream (ATM switches) and Kalpana (legacy switches). Acquiring Grand Junction has provided Cisco with expertise in high-speed fast Ethernet networks. The real key to the company's success is likely to come from how well these different products can be integrated with Cisco's routers and with each other. The company's ability to create a seamless network with complete interoperability will determine its success in the marketplace. In this chapter, I'll examine Cisco's switches as well as its specific architecture—CiscoFusion—for the way the company believes networks will develop. I'll also describe Cisco's plans for distributing routing functionality throughout a network while preserving the investment of companies that have purchased large Cisco backbone routers.

CiscoFusion Architecture

Cisco touts its *CiscoFusion* architecture as a roadmap to help companies build switched internetworks to handle high-bandwidth applications and also shared-bandwidth legacy applications. The architecture promises support for all desktop platforms, operating systems, applications, and associated data protocols. It's particularly appropriate for multimedia applications.

Given Cisco's history as a router company, it should be no surprise that routing plays a major role in CiscoFusion. CiscoFusion architecture includes router clusters, multiple routers connected to an ATM switch, and in the future perhaps switches with some routing functions built into them. The architecture also includes ATM as a core data-switching engine with legacy LAN switches used for both LANs and desktop power users.

Cisco believes that future networks will need to have some switches that operate at layer 3 of the OSI model, while other switches will need to support only the layer 2 protocols already supported by today's switches. The creation of virtual LANs are an integral part of CiscoFusion architecture. Cisco sees a problem with today's switches that operate at layer 3 because the switching decisions made at that layer require the switch to assemble a table of routing information. That requires CPU cycles and adds cost and administrative complexity. Cisco's solution is to separate layer-3 switching from route calculation. This task is performed by one or more route processors that reside in either the routers or ATM switches. These devices periodically distribute path information to the multilayer LAN switches. Cisco's plan is to have its multilayer switches store layer-3 information in cache memory and refer to the appropriate path information when needed to forward packets to a local destination. The result is that these switches won't need to send a packet through a router for route calculation.

Cisco argues that its 7000 family of high-end routers already supports an architecture with three components: a route processor to calculate routes, a switch processor to make packet-switching decisions, and a number of interface processors for input from and output to connected network segments.

Figure 8.1 illustrates what Cisco calls its *switched internetwork model* for CiscoFusion. Throughout this chapter I'll discuss each of the key components in some detail because the company's success or failure will be largely based on how well customers accept the concepts inherent in this blueprint for the future.

How Switching Will Evolve

Cisco believes that network switching will evolve in four distinct phases. The first phase, which it refers to as *microsegmentation*, is characterized by companies retaining their hubs and routers but enhancing performance by adding a LAN switch. The second phase adds ATM technology and routing between switches, and LAN switches are found in some abundance. Backbone routers are clustered by ATM switches to optimize backbone bandwidth. The third phase connects the ATM core switches directly to the LAN switches as well as to centralized or distributed ATM routers. At this phase a network can be characterized as an ATM-centric because the backbone is composed of ATM switches, with all other devices at the periphery. The network also includes mul-

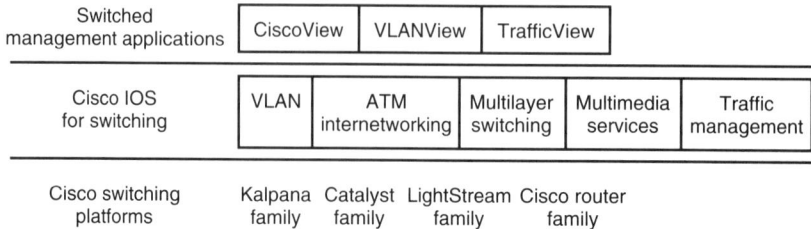

Figure 8.1 CiscoFusion's switched internetwork model.

tilayer switches that can switch at layer 2 and layer 3 or a combination of both. The final phase is end-to-end switching with integral virtual LAN and multilayer switching capability. Route and switch processors are distributed over the ATM fabric. If the future evolves in this fashion, clearly Cisco will be very well positioned to be a major player.

Cisco Pushes a VLAN Standard

Cisco has used the IEEE 802.10 Committee as a forum to push its initiative for a standard for virtual LANs. While this committee was originally created to design specifications for security within LANs and metropolitan area networks (MANs), Cisco proposes that a four-byte field in the 802.10 frame be used to help hubs and routers to identify network traffic by their VLAN information. At this point several vendors are resisting this initiative because they see it as an effort by Cisco to enhance its market position. Right now each vendor has its own proprietary method to handle VLAN traffic.

The Cisco Internetwork Operating System (IOS)

The *Cisco internetwork operating system (IOS)* is Cisco's software that runs on a number of different network platforms (NetWare, Windows NT, etc.) that makes them appear to be a single integrated whole. Cisco is placing this software on stand-alone routers, router modules in hubs, and stackable hubs. It's also placing it in PC and workstation file servers, X.25 switches, ATM switches, and ATM-capable PBXs. Among the IOS functions are monitoring a network's logical structure and managing, controlling, and logically routing traffic. It also provides firewalls, gateways, filtering, and protocol translation. The IOS for switching software uses LAN emulation protocols so virtual LANs can be created across a network that includes ATM switches. LAN emulation configuration servers will first be implemented on Cisco routers but then later on various switching systems in Cisco's product family.

IOS and Multimedia

Cisco intends to be a major player in ATM, and it sees multimedia as a key application that must be supported under IOS. It sees three key elements to multimedia:

- Guaranteed end-to-end quality of service
- Multicast packet delivery for efficient bandwidth usage
- Scalable bandwidth to support variable multimedia requirements

The company is incorporating a weighted fair queuing algorithm into IOS to reduce latency and minimize jitter, two conditions that can degrade the quality of service associated with desktop video conferencing. *Weighted fair queuing* enables a network to identify various types of network traffic and assign them consistent routing prioritized to minimize latency.

Cisco is also working with the Internet Engineering Task Force (IETF) to gain acceptance for the *resource reservation protocol (RSVP)*. This is an internetwork protocol that allows an application to dynamically reserve resources for different classes of service using the capabilities of the underlying network (such as ATM).

Cisco supports IP multicast multiprotocol packet delivery on both router and switch platforms. A computer transmits a packet addressed to all intended recipients and, under IP multicast, the network replicates the packet only when necessary. Cisco ships IP multicast with its IOS.

Finally, Cisco believes it's providing users with solutions to their variable amounts of bandwidth by offering a variety of different products. It points to its Kalpana and Catalyst workgroup switches as effective means for segmenting LANs, and to its FDDI and ATM routing and switching as ways of increasing backbone bandwidth.

Legacy LAN Switches

Cisco's purchase of Kalpana gave it the legacy LAN switch component of its CiscoFusion architecture. A look at one key product will illustrate some of the key features of this switching family. The ProStack system is a rack-and-stack stackable switch with 4.8 Gbps of bandwidth that can be stacked up to eight units. It can operate at 10 Mbps, 100 Mbps, and 155-Mbps ATM. It also includes a ProStack 100 Base-TX fast Ethernet module and the ProStack matrix, an eight-port, nonblocking, cross-int switch for connecting multiple EtherSwitch Pro 16s. It also includes the ProStack port, an expansion module that provides connectivity to the ProStack matrix. Two expansion slots can be used for either ATM or 100 Mbps. It supports 100Base-T topology and will eventually support 100VG-AnyLAN topology. Cisco has made it clear that it won't become involved in the wars involving which 100-Mbps Ethernet standard will prevail; they'll simply let the market decide by supporting both. There are no wide area network ports on the ProStack/Pro16 because ATM switches are expected to play this role.

The Cisco Catalyst 5000 is a modular, chassis-based switching hub with five slots. One slot is dedicated to a network management module that comes with an SNMP agent and embedded RMON. The hub supports two fast Ethernet ports as well as switched Ethernet, 100Base-T fast Ethernet, Token Ring, 100VG-AnyLAN, ATM, and FDDI. The Catalyst 5000 architecture supports multiple levels of data prioritization with each interface separately user-configurable with high or low priority. Because separate logical queues for each priority class are maintained, there's a marked reduction in buffering delays; this makes the Catalyst 5000 ideal for bursty network traffic and delay-sensitive traffic such as voice and video. Cisco has identified the Catalyst 5000 as its platform for multilayer switching, with the ability to switch packets at both layers 2 and 3.

LightStream 100

The LightStream 100 is a high-performance modular workgroup/campus ATM switch. It offers 16 ATM interfaces. Among the interfaces supported are the following:

- 155-Mbps SONET/SDH OC3 over single- and multimode fiber
- 155-Mbps STS3c over unshielded twisted-pair wire
- 45-Mbps DS3

- 34-Mbps E-3 over coaxial cable
- 100-Mbps TAXI 4B/5B

The switch supports both permanent virtual circuits (PVCs) and switched virtual circuits (SVCs). It should come as no surprise that this switch operates seamlessly with the Cisco 7000 family of high-end routers, a link that Cisco believes makes it easy to create switching internetworks.

The switch has a nonblocking 2.4-Gbps switching fabric that can handle a minimum of 1,000 virtual output cell buffers per port. It provides up to two priority levels to cover both cell loss and cell delay parameters. A key software feature is *permanent virtual path (PVP) tunneling*, which enables a network manager to set up temporary switched virtual connections over previously established PVPs. This approach makes it possible to set up an ATM network signal through a carrier's ATM service even if it offers only permanent connections. Sometimes called "soft" permanent virtual circuits, the PVP feature lets network managers set up permanent connections across multiswitch networks by designating the two end points of the connection, a much faster approach than having to manually create connections between all devices along the connection path.

Cisco believes the primary role of the LightStream 100 will be to construct campus backbone networks that connect a number of ATM routers, switches, and high-performance servers into what the company calls a *router cluster*. It defines an ATM router as a member of the Cisco 7000 family, with an ATM interface processor (AIP) module. This module permits LANs to be interconnected across ATM backbones. It also paves the way for the creation of virtual networks. Figure 8.2 shows Cisco's vision for ATM networks.

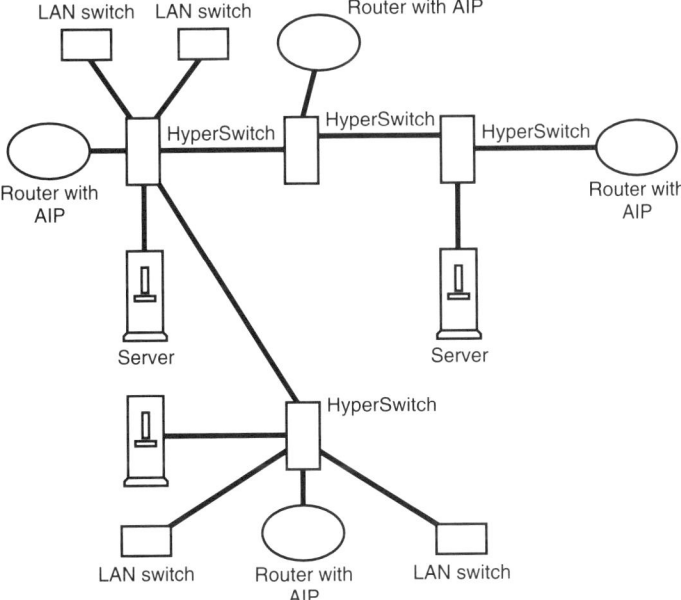

Figure 8.2 Cisco's vision of ATM networks.

The Cisco internetwork operating system for the switch includes dynamic IISP and ILMI, features that automatically set up connection paths and map addresses respectively. *Dynamic IISP* is a dynamic version of the ATM Forum's interim interswitch signaling protocol. It allows the switch to automatically establish connections and pass "reachability" data to eliminate the need for manual specification of connection paths. The software is smart enough to dynamically reroute traffic around failed devices. This maximizes network reliability. The *interim local management interface (ILMI)* is an ATM Forum standard that enables an ATM end station to register its network address automatically with an ATM switch. This means that network managers don't have to map addresses manually.

The LightStream 100 supports the ATM Forum's standard point-to-multipoint signaling, so it can automatically create unidirectional one-to-many switched virtual circuits. This feature is particularly useful for improving bandwidth usage in video broadcasts.

LightStream 2020

The LightStream 2020 supports T-1 circuit emulation, so it can be connected to a PBX or video conferencing equipment. The Cisco internetwork operating system software provides quality-of-service features (application-specific quality of service, AS/QoS) to prioritize bandwidth. The major advantage of this support for T-1 emulation is that it makes it easier for network managers to justify purchasing ATM switches because they can carry voice and video in addition to the data. AS/QoS extends ATM quality of service to FDDI and Ethernet traffic. There's also a frame-relay interface that connects to other vendors' frame-relay switches. A topology map enables the network manager to view network status via a color-coded map that shows traffic types and priorities as well as changes in physical topology.

High-performance multicast service (HPMS) enables the 2020 to forward multicast packets at full line rate within the ATM infrastructure. It permits the user to allocate bandwidth and priority of service by applying it to multicast groups using point-to-multipoint ATM connections.

Internetwork Management Software

Cisco offers a number of software tools to help network managers manage their virtual LANs. VLANView is an application with a graphical user interface that can be used to configure networks by logical user groups and display a physical view of each Cisco device on the network. It can also assign virtual memberships for each switch port, display and print topology maps for each VLAN, and provide reports with all physical port assignments.

TrafficView is a tool for monitoring and analyzing traffic. It can access information from the embedded RMON agents found in Cisco switches. Data can be captured for offline analysis in order to track protocol-related problems. It can display and print reports on network usage and growth.

CiscoView

The CiscoView software provides a graphical view of the physical state of the device being managed as well as all interfaces on that device. Users can configure the device, monitor performance, and perform some minor troubleshooting functions. The physical devices are displayed in real time. Users can click on a device and see detailed information on its status.

Should Cisco Systems Be Your ATM Vendor?

Cisco has come a long way very quickly by filling out its product lines with some very successful acquisitions. It now offers hubs and switches as well as routers, and plans to embed much of a router's functionality in silicon so the functions can be distributed throughout a network. Some critics have pointed out inefficiencies in this approach to routing. CiscoFusion requires as many as four protocols to pass routing instructions. A Cisco multilayer switch must query a route server. Because this switch doesn't support the ATM Forum's PNNI standard, it uses a protocol such as IGRP to initiate contact with the route server. It then uses the next hop resolution protocol (NHRP) to ask for a route. This protocol lacks quality-of-service parameters so latency, cell delay, and bandwidth requirements can't be specified. If two route servers need to check with each other across the enterprise, they use the route distribution protocol to communicate.

Critics believe that Cisco's use of all these protocols is inefficient and comes from the company's overemphasis on maintaining the importance of routing in future switched networks rather than consigning routing to the periphery in the form of edge routers or building routing functionality into switches. Having pointed this out, however, companies with significant investments in Cisco 7000 backbone routers can certainly leverage these investments by looking to Cisco as a partner for future migration to ATM.

There are two major limitations that might cause some potential customers to look for an alternate vendor when selecting ATM equipment. One is that Cisco Systems has acquired so much technology in such a short period of time that some potential customers might be concerned about how well integrated these products are at the present time. A second limitation is that Cisco's desire to embed so much of the routing functionality in silicon and then to distribute this function as hardware is bound to create interoperability problems with other vendors' products. Cisco's proprietary handling of virtual LANs creates a difficult situation for companies that might already have purchased legacy switches from other vendors. Can they count on Cisco to make everything work together smoothly? The jury is still out.

Summary

Cisco Systems has been acquiring companies to complement its own router product line. By adding an ATM switching company (LightStream), a legacy LAN switching company (Kalpana), and a company focused on fast Ethernet (Grand Junction),

Cisco has assembled the building blocks to offer a complete network switching solution. The key blueprint that Cisco has unveiled to describe its vision of a switched network is CiscoFusion. Legacy LANs, routers, legacy LAN switches, ATM switches, and hubs are all key components of CiscoFusion. These devices will be running Cisco's internetwork operating system (IOS) software.

The company sees several stages of network evolution, reaching culmination in an all-switched network in which hubs, routers, and servers are all on the periphery of the network. It sees the routing function being distributed throughout a network, and the development of multilayered switches that can function at both layers 2 and 3 of the OSI model.

9

Major Hub
Vendors and ATM

How will ATM be implemented in enterprise network environments? Several major intelligent hub vendors are banking on hub-based visions of ATM implementations that include hubs with ATM backplanes supplemented with ATM and legacy LAN switching modules. In the case of Bay Networks and 3Com, both vendors have filled in gaps in their switched enterprise network offerings by acquiring companies with the necessary technology. Even Cabletron, a company that usually prefers to develop its own products, has opted for a strategic partnership with FORE Systems to help it fill in some key pieces to the switched network puzzle. The major hub vendors all have large installed bases of customers with significant investments in Ethernet and Token Ring equipment. Both 3Com and UB Networks have developed intriguing ATM migration paths for their customers with legacy LAN equipment. Cabletron has even developed a way to penetrate its competitors' customer bases by positioning itself as the one company that can manage mixed-vendor environments.

In this chapter I'll focus on the products, product positioning, and overall ATM strategy and architecture offered by Bay Networks, 3Com, Cabletron Systems, and UB Networks.

Bay Networks

The "as equals" merger of hub vendor Synoptics and router vendor Wellfleet created the magacompany Bay Networks. Bay, in turn, acquired Centillion Networks to gain access to their Token Ring and ATM switching technology. The company has developed *BaySIS* (Bay Networks switched internetworking services) as its architecture to migrate customers from their present shared-media network environment to a future switched internetworking environment. This architecture covers the entire enterprise, from backbone to desktop, and even encompasses remote site access. It's built on three fundamental services that a network must provide: transport, operations, and policy services.

Transport services

The BaySIS philosophy for transport services is to "switch where you can and route where you must." Transport services move data, video, and voice traffic through a switched internetwork. Bay Systems breaks sharply with Cisco and IBM in its position on the role of routing in a switched network environment. The company plans to integrate routing and switching in its products, particularly those deployed in the LAN backbone. It will introduce a proprietary extension to the PNNI standard, known as IPNNI. This set of proprietary interfaces will map routing data from protocols such as IPX and DECnet for transmission across PNNI links. It's the way Bay Systems intends to implement the ATM Forum's multiprotocol over ATM standard. This extension enables a router to transmit routing information across an ATM network to other routers. It even permits these routers to extend ATM quality-of-service parameters to legacy LAN systems. BaySIS will support Token Ring switching and APPN HPR, IBM's ATM-ready network protocol. All routers in such an environment would share a common view of the network's topology as well as the cost associated with various paths. Figure 9.1 illustrates the role of IPNNI.

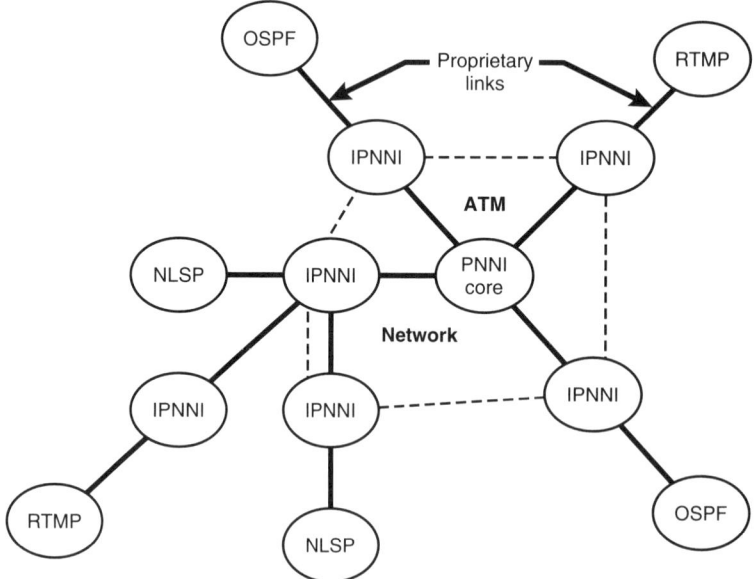

RTMP = Routing table maintenance protocol
NLSP = NetWare link state protocol
IPNNI = Integrated PNNI
PNNI = Private network-to-network interface
OSPF = Open shortest path first

Figure 9.1 The role of IPNNI in a switched network environment.

Policy services

Network managers need the flexibility to create and manage secure virtual networks. BaySIS extends current virtual network support from the MAC layer to the network layer so networks can be created based on a common protocol or subnet. It also supports extending these virtual networks across wide area networks. In order to manage these widely dispersed networks, BaySIS supports network-wide traffic management for router-based networks, as well as centralized remote office configuration and control for these networks.

Operation services

BaySIS's blueprint for future switched internetworks includes the following support for day-to-day monitoring and control of the network:

- Asset management
- Network configuration
- Analysis, planning, and design
- Accounting and billing
- Troubleshooting and fault management

The Bay Networks 5000 Hub

For Bay Systems, the focus of a network is its enterprise hub, the SynOptics System 5000. This hub efficiently manages multiple Ethernet, Token Ring, and FDDI networks. The 5000AH switch for the 5000 is a 16-port LattisCell ATM switch with an integrated ATM control module running connection management, network management, and LAN emulation software. It has an aggregate switching capacity of 5 Gbps and supports both single-mode and multimode SONET fiber interfaces. Future plans call for adding an ATM backplane to the 5000 to enable it to integrate Ethernet-to-ATM and Token-Ring-to-ATM switching modules.

While BaySIS plans call for eventually integrating routing functionality in switches, Bay Networks has to deal with current customers with the SynOptics System 5000 hub who want to route data over ATM links. The solution is what it calls *virtual network routing (VNR)* using the ATM routing engine (ARE) for the Wellfleet backbone node router. It provides 155-Mbps, full-duplex, ATM routing between multiple virtual LANs. In a collapsed virtual internetwork backbone architecture, the ARE connects the backbone node to the Lattis System 5000 switch for the System 5000, and performs routing for all segments that have downlinks to the 5000.

ARE can take packets from a remote LAN and convert them to cells before transmitting them to the 5000. It can also take cells from the 5000, repackage them as packets, and then transmit them to WAN ports on the backbone node for transmission to wide area network LANs. Finally, ARE can function as a LAN emulation client to an ATM network. It can take LAN packets from distant LANs, request an ATM address for the packets from the 5000 (acting as a LAN emulation server), and then convert the cells before transmitting them to the appropriate VLAN over the appro-

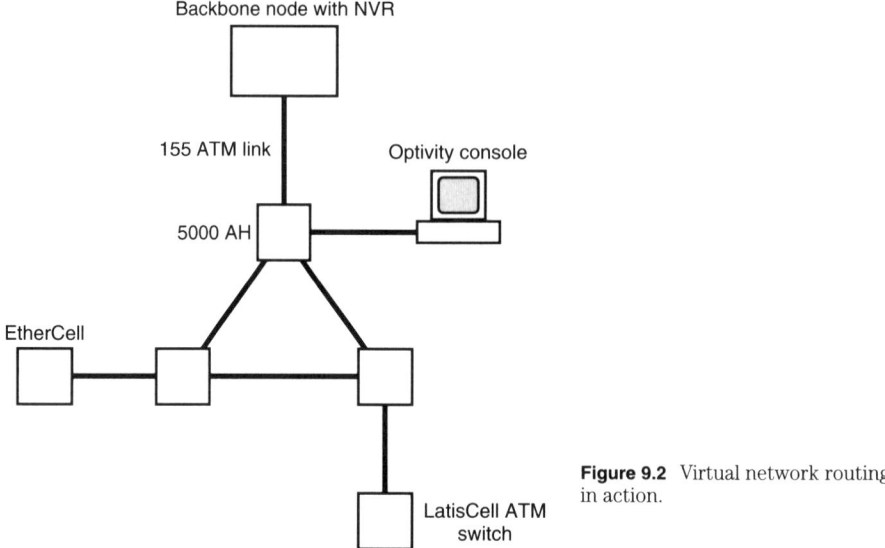

Backbone node with NVR

155 ATM link

Optivity console

5000 AH

EtherCell

Figure 9.2 Virtual network routing in action.

LatisCell ATM switch

priate ATM circuit. All traffic can be managed via an optivity console attached to the network. Figure 9.2 illustrates virtual network routing in action.

Bay Networks switches

The LattisCell ATM switches share a number of features, including their large output buffers, dual output queues, and flash memory. The output buffers on each of the 16 LattisCell ports can hold up to 1,024 cells. This means that each output port can support multiple bursty communication streams. Each output buffer offers dual queues based on high priority and normal priority traffic. The LattisCell switches can support multiple traffic types and the five quality-of-service classes outlined by the ATM Forum's UNI 3.0 specifications. At this time, LattisCell switches support 155-Mbps desktop connections and not ATM 25 connections.

Bay also offers an EtherCell switch that translates Ethernet frames into ATM cells. This preconfigured switch multiplexes 12 10Base-T ports onto one ATM-Forum-compatible 155-Mbps SONET/SDH port. This switch is targeted at two specific applications: power workgroups and backbone networks.

Each 10Base-T port provides a full 10-Mbps dedicated bandwidth, while the ATM port provides access to the ATM network fabric with an aggregate bandwidth of 5 Gbps. Both the EtherCell and LattisCell switches can be administered through Bay Networks connection management system. This software resides on a Sun SPARC-station. It establishes virtual channels on a point-to-point or point-to-multipoint basis. It can discover the network topology and location of all switches and nodes on the network and determine the optimum data paths required. It includes the multicast server software that supports broadcast and multicast services based on the ATM Forum's LAN emulation specifications. The connection management system software acts as a simple network management protocol (SNMP) proxy agent for a network of Bay Networks switches. The data is transmitted to a management station running the Bay Networks' ATM management application.

Bay Networks' acquisition of Centillion offers the speed switch 100. This is an ATM core LAN switch with 10 Gbps of aggregate switching capacity. A Token Ring module added to the switch provides four-wire, speed-switched, Token Ring ports. Users can also add an ATMSpeed/155 switch module. This module comes in two- or four-port OC-3 versions. It allows users to transport LAN packets as streams of cells over the backbone. What's intriguing about this switch is that it enables network managers to upgrade select users to ATM by simply adding modules where they're needed. An FDDI switch module (FDDISpeed) can also be added. The speed switch network management software supports an SNMP agent and an IBM LAN network management agent. SpeedView network management configuration and monitoring applications operate in a number of different environments, including NetView/6000, OpenView, and Windows.

Should Bay Networks be your ATM vendor?

Bay Networks is a particularly attractive ATM vendor if your company has already invested in SynOptics or Wellfleet equipment. One problem some companies might have with Bay Networks is that it offers some proprietary solutions. A second problem is that, while Bay Networks was one of the first vendors to offer a detailed description of its plans for ATM migration, products have been slow to appear and a complete solution is not yet in place.

3Com

3Com's growth has been phenomenal. Much of it has come through the acquisition of other companies. The company merged with Bridge Communications for its key routing technology, and its President (Eric Benhamou) eventually became 3Com's CEO. The company acquired BICC for its Ethernet hubs, Star-Tek for its Token Ring hubs, switching technology from Syneretics, ATM technology from NiceCom, remote technology from Centrum Communications, and Chipcom for its channel connections. What's amazing is that the company has managed to integrate its acquired product lines and offer a coherent network architecture plan as well as a common network management software application (Transcend). Benhamou has indicated that he isn't yet finished acquiring companies.

Benhamou sees switching as the new network model that flattens networks by eliminating routing boundaries and creating virtual networks that are more flexible and easier to manage and configure. He also believes that networks formed of large, medium-size, and smaller switches will help break the scaling bottleneck created by the need to centralize configuration and routing. He foresees a time in the near future when companies will have ATM in their backbones, Ethernet on the desktop, and a fast Ethernet connection to the backbone. The company's initial plans for desktop ATM consists of an Sbus adapter and a PCI bus adapter. The company believes that ATM won't be a factor on the desktop for several years, but will be a factor in server farms much sooner, so it will focus on server-oriented adapter products.

High-performance scalable networking (HPSN)

3Com has developed *high-performance scalable networking (HPSN)* as its strategy for providing high-speed links for growing networks as well as networks running advanced applications. The company has promised that HPSN will deliver:

- Increased performance and manageability at every level of the network
- Economical upgrades using network platforms and architectures that are available today
- Routed ATM, combining the speed of ATM with the traffic control of routing

In addition, 3Com has promised to "future-proof" its network components so existing LANs will always be supported. The company has also developed guidelines so its customers can accomplish their entire migration process around two of its key platforms: the NetBuilder bridge/router and the LinkBuilder multiservices hub. 3Com has promised that HPSN, which is built on the concept of collapsed backbones, will take place in three phases. 3Com is a great believer in the benefits of *collapsed backbones* over *distributed backbones*. Distributed backbones in multistory buildings would consist of LAN segments on each floor connected to a repeater in the basement through routers. A collapsed backbone eliminates the need for a router on each floor. It concentrates all connections in a single device in the basement. The distributed backbone is "collapsed" onto the high-speed backplane of a router. Hubs concentrate the LAN segments on each floor, but the intelligence and complexity of the network are found in the basement with the router. Figure 9.3 illustrates 3Com's concept of a router-based collapsed-backbone architecture.

3Com says there are several reasons why a router-based collapsed backbone architecture is more efficient than a distributed backbone approach. The need for only one router port per floor instead of an entire router obviously saves money. The company also argues that network management is more efficient and easier and that there's improved scalability. This architecture serves as the foundation of 3Com's HPSN.

HPSN stage 1: collapsed backbone with multiple LANs. HPSN's first stage consists of collapsed backbones with multiple LANs. 3Com believes that companies handle the problem of performance bottlenecks by segmenting a LAN on each floor in a multifloor building into multiple LAN segments. By adding downlinks for each LAN or LAN segment between an individual floor's hub and the router, bandwidth is increased in direct proportion to the number of downlinks. In other words, if the LAN on the top floor is segmented within the hub on that floor into four LAN segments, and each segment is connected via a downlink to the router in the basement, then the bandwidth on the top floor in effect is quadrupled. The major limitation of this approach is in the finite number of router ports available. A second limitation is the growing complexity of the LAN environment as LAN segments are added, each with their own network address.

3Com addresses the problems found in stage 1 of HPSN by offering a significant amount of segmentation in its LinkBuilder MSH as well as its LinkBuilder FMS stackable hubs. The company has doubled the amount of LAN ports on its NetBuilder bridge/router to facilitate additional downlink connections.

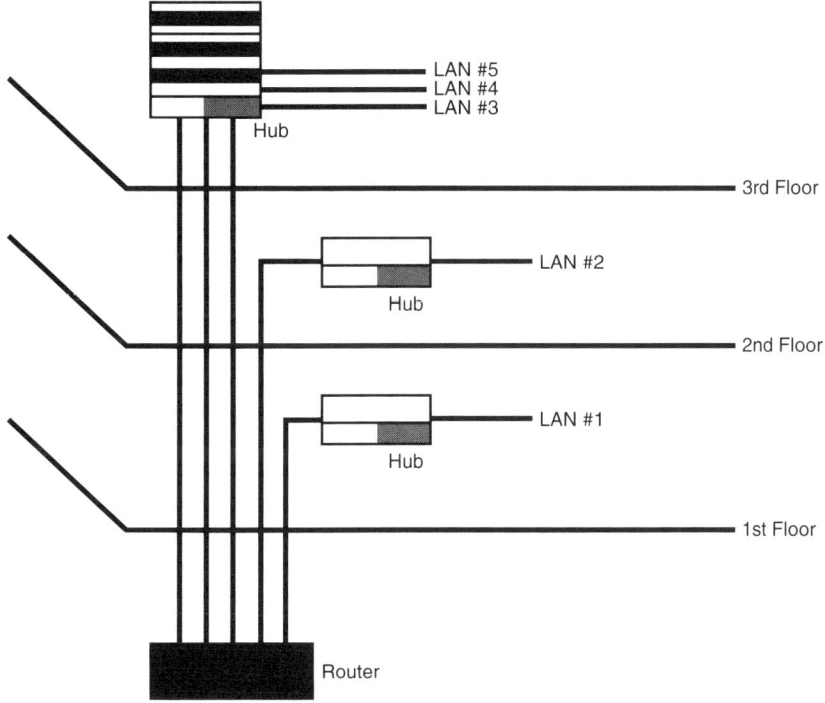

Figure 9.3 3Com's router-based collapsed backbone architecture.

HPSN stage 2: collapsed backbone with ATM downlink. This stage of HPSN replaces the slower downlinks found in stage 1 with a single, high-speed, 155-Mbps, ATM link. A single ATM link can support up to 30 Ethernet LANs or 20 16-Mbps Token Ring LANs on each floor, according to 3Com. Because each LAN segment can maintain a different virtual connection within the downlink, each segment is identified by the router so the network manager can create virtual workgroups just as if all the slower downlinks were still present. 3Com's current implementation of stage 2 consists of a CellBuilder module within the LinkBuilder MSH hub that converts Ethernet and Token Ring traffic into ATM cells. This module can also receive ATM cells from the downlink and reassemble them into LAN packets for transmission to the stations on the network. 3Com plans to extend the software in NetBuilder II so each virtual connection in the ATM downlink can be treated by the bridge/router as a logical port. This way, all the bridging and routing associated with a physical port applies to each logical port.

Because each virtual segment and thus each LAN segment can be directed to a specific router or router port, it's possible to use load sharing across the routers to optimize traffic patterns. The load sharing can be split across routers and hubs to create some system fault tolerance, and can be accomplished through software without any physical changes in the network.

HPSN stage 3: collapsed backbone with routed ATM. The logical growth of networks at stage 2 will ultimately create a bottleneck within the backbone. To overcome this problem, Stage 3 introduces a router in front of the ATM switch and route caching. To 3Com, *route caching* is using a high-speed memory cache on the interface module to hold forwarding information. When a bridge/router receives a request for routing, the main routing engine decides how to forward the information and saves the data associated with its decision to forward to a high-speed memory cache on the interface module. The advantage of this approach is that when subsequent packets have the same destination network address, the routing information is available immediately.

3Com has committed to managing all stages of HPSN via its Transcend object-oriented network management application. Its network architecture permits use of the company's current products and SmartAgent device agents, and supports SNMP-managed products from other vendors. 3Com expects its future network management applications to run on the most popular platforms, including SunNet Manager, OpenView/Unix, and NetView/6000.

3Com leverages its chip technology

3Com plans to leverage its expertise in application-specific integrated circuit (ASIC) development on three different platforms. Its intelligent switching engine (ISE) chip has been developed for Ethernet and FDDI networks and will be a part of the company's LANplex 2000 switching hubs. Its ZipChip1 chip has been developed for ATM networks, while its Brasica chip has been developed for Ethernet networks.

The 3Com product line

3Com argues that it offers "a switch for every technology and network level." Its switches support Ethernet, 10Base-T fast Ethernet, Token Ring, FDDI, and ATM. All switches are managed from the company's Transcend network management software or via an SNMP-based third-party application. 3Com has created three different product areas for its switching architecture. The LANplex family includes LAN switches for workgroups, departments, and data centers. These switches use a technique known as *elastic packet buffering* to ease port congestion by dynamically allocating additional buffers as needed. The LinkSwitch family includes stackable LAN switches and SuperStack software modules for the LinkBuilder MSH hubs. This means that these switches can be combined into a SuperStack system to increase port density and add management functions. Finally, the CellPlex family consists of modular, nonblocking, ATM switches.

3Com supports ATM-Forum-standard LAN emulation on its backbone and workgroup switches as well as on its network interface cards so legacy LAN applications can be run over ATM. It believes it offers the most comprehensive line of products that support standard ATM emulation, as well as a smooth migration path to ATM. The company has placed its LAN emulation software in the CellPlex 7000 campus backbone switch, the LinkSwitch 2700 Ethernet/ATM workgroup switch, the CellPlex 7200 Ethernet/ATM switch, the MSH Ethernet/ATM switch, the LinkSwitch 3000 fast Ethernet/ATM switch, and the LinkSwitch TR Token Ring/ATM switch. 3Com's Transcend network management software supports virtual LAN management on ATM networks.

The LinkBuilder MSH multiservices hub

3Com's vision of enterprise networks includes a great deal of emphasis on its LinkBuilder MSH multiservices hub. This product can support up to 10 managed Ethernet, 5 managed Token Ring, or 3 managed FDDI workgroups. 3Com plans to implement fast Ethernet and ATM in this hub. The hub can be managed with the company's Transcend management application or any SNMP-based application.

Legacy LAN switches

LANplex switches are designed for collapsed backbones, data center, departments, server farms, and workgroups. The LANplex 6000 series offers Ethernet and Token Ring switches as well as switched FDDI, FDDI concentration, bridging between technologies, and intranetwork routing. Because each module comes with its own processor, performance scales up with each module added. The 19.5-Gbps backplane makes the LANplex 6000 an ideal high-speed collapsed backbone. A built-in FDDI connection links this switch to other backbones or switching hubs. 3Com plans to add fast Ethernet and ATM modules in the future. An ATM uplink will enable this hub to connect to ATM switches from other vendors, such as FORE Systems.

LANplex switches also support virtual LANs. Managers have three different configuration options for virtual LANs with these switches. They can select port grouping to define broadcast domains and control traffic by creating groups of ports. They can select MAC address grouping to restrict traffic down to the workstation level. Finally, they can select IP routing to define virtual LANs by the existing IP subnet structure.

ATM products

The LinkSwitch 2700 is an Ethernet/ATM workgroup switch that has 12 Ethernet ports (10 Mbps each) and one ATM port (OC3c, 1500-Mbps SONET/SDH). It's positioned as an ideal switch for workgroups and small departmental LANs that require a high-speed ATM downlink to an ATM campus backbone. The switch has software-selectable modules available, in both the cut-through and store-and-forward varieties.

The ATM port need not be configured for the switch to be used. This feature means that companies can use the product for Ethernet switching right now and then add interoperability to an ATM backbone at a later date. The switch is manageable via 3Com's own Transcend application or via SNMP-based applications from other vendors.

3Com offers its CellPlex 7000 campus backbone switch as well as its LinkSwitch 2700 Ethernet/ATM workgroup switch. The CellPlex 7000 is a nonblocking switch with a 2.5-Gbps switching capacity. The switching engine features a cut-through, self-routing architecture built around a 20.48-Gbps backplane. Each port can support up to 4,096 virtual-channel connections that can be both point-to-point and point-to-multipoint. There's a separate on-board I960 processor to handle all advanced software features, including switched virtual-circuit signaling, LAN emulation servers, and SNMP management.

The CellPlex 7000 supports from 1 to 16 ATM ports, depending upon how many four-port interface cards are installed. Each card handles up to four OC-3c 155-Mbps SONET/SDH interface modules or up to four DS-3 45-Mbps interface modules for WAN connections. The switch is fully redundant with dual power supplies, redundant switching engine, and no single point of failure. Modules can be hot-swapped for continuous operation.

The CellPlex 7200 switching hub provides Ethernet and ATM switching and is designed to ease traffic congestion in backbones and departmental LANs. It supports up to 48 switched Ethernet ports and four ATM ports, or up to eight ATM ports in an ATM-only configuration. The ATM ports handle OC-3c 155-Mbps SONET/SDH interfaces as well as DS-3 45-Mbps WAN ports. The switch's passive backplane has a 10.24-Gbps capacity.

The NetBuilder II bridge/router

Very conspicuous in 3Com's plans for its high-performance scalable networking (HPSN) is the role of the NetBuilder II bridge/router. It integrates Token Ring, Ethernet, FDDI, ATM, and WAN connections. The product supports all major LAN and WAN protocols, including TCP/IP, IPX, XNS, TM OSI, DECnet phase IV and V, VINES, AppleTalk phase II, IBM DLSw, APPN, ISDN, X.25, frame relay, SMDS, and ATM. NetBuilder offers a multiprocessor (MP) architecture that integrates the processing capabilities of MP modules and the NetBuilder II's communications engine card (CEC). The CEC offloads most filtering and forwarding operations to the MP module's onboard processors for greater efficiency. FORE Systems is developing an ATM card for this product.

Routing or switching?

3Com thinks the choice in most companies today is not between routing or switching. It believes that about 80 percent of 3Com networks have a router and LANplex switch sitting side by side in the basement to provide a network's collapsed backbone. When companies choose between the LANplex 6000 and the NetBuilder II, 3Com believes the decision is often based on political and not technical grounds. The router is also chosen sometimes because it can prevent access to certain resources.

3Com's ATM products in action

3Com has pointed to Tel Aviv University as an early adopter of its ATM technology. This campus, with 23,000 students and 50 buildings in a campus environment of 220 acres, uses an ATM backbone. This backbone is expected to carry multimedia and information browsing applications, as well as video on demand and teleconferencing and several different scientific applications. A 3Com CellPlex 7000 switch and LinkSwitch 2700 workgroup switches link the School of Mathematical Sciences with the University's Computer Center. The various platforms supported at the school include Ethernet and ATM, as well as Sun SPARC10 and Silicon Graphics Indigo workstations and PCs. The CellPlex backbone switch is installed in the Computing Center

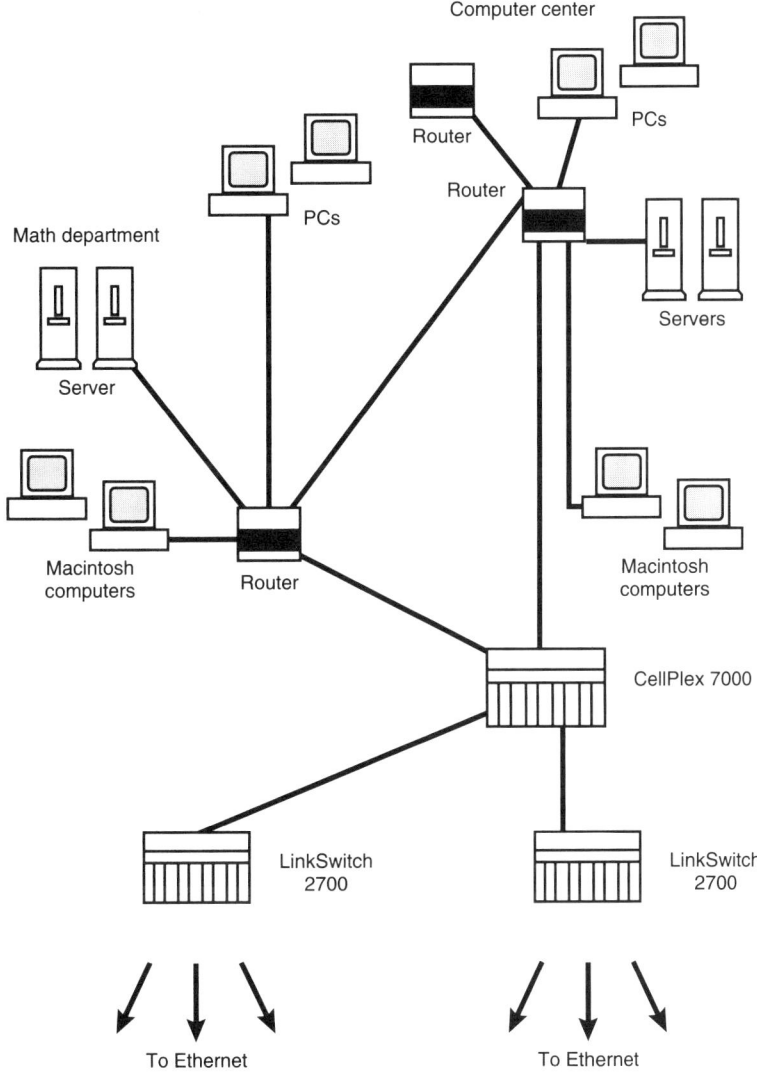

Figure 9.4 3Com's ATM network implemented at Tel Aviv University.

and linked directly to two LinkSwitch 2700 LAN access switches, which provide ports for Ethernet LAN segments or dedicated Ethernet workstations.

The school's campus ATM networking is evolving. They expect to incorporate some ATM equipment from both FORE Systems and Cisco Systems. Figure 9.4 shows the current campus ATM environment.

Should 3Com be your ATM vendor?

3Com has done a remarkable job during the past few years integrating the technology from the companies it has acquired. So far it has fulfilled its promise to its customers that its ATM and switched network migration strategy would protect their equipment investments. Companies with heavily mixed protocol environments might be attracted to 3Com because of its distinctive router-based ATM strategy and its support of all major protocols. Companies without a current investment in large routers, without so many protocols to support, and without current 3Com equipment, however, might prefer the distributed routing approach taken by other vendors.

Cabletron

Cabletron has long been an adherent of going it alone with minimum help from partners. It has been quick to point out the significant sums of money it's able to spend on research and development as a plus for its customers. Unlike 3Com and Bay Networks, Cabletron has not chosen to acquire companies to gain technology. As you'll learn in this section, Cabletron has developed a strategic relationship with FORE Systems while at the same time building it's own comprehensive switching network architecture.

Legacy switching and accessing ATM

Cabletron offers a variety of switching options, with an emphasis on Ethernet and FDDI. *The multimedia access center (MMAC-Plus)* is an enterprise network platform, the *flexible network bus (FNB)* model is designed for wiring closets, and the *internal network bus (INB)* platform is designed for data center applications.

Cabletron and FORE Systems

Cabletron has a strategic partnership with FORE Systems. The ATM switching module for its MMAC-Plus enterprise hub was codeveloped with FORE. The module is based on FORE's ASX-200 workgroup switch, but has been modified to use the Intel I960 chip found in all Cabletron devices. The switching module features a 2.5-Gbps nonblocking backplane, and each module supports as many as 24 ports. The switching module can support a number of interfaces, including TAXI (100 Mbps), SONET (155 Mbps), DS3 (45 Mbps), DS1 (1.544 Mbps), and E1 (2 Mbps). It also supports ATM Forum UNI 3.0 as well as switched virtual circuits. One MMAC-Plus hub can hold seven modules.

Cabletron's ATM access module provides ATM access to the MMAC Plus's flexible network bus (FNB). This module translates LAN packets into ATM cells and runs at 155 Mbps. It supports both private virtual circuits and switched virtual circuits. The company's LAN emulation software is based on FORE's simple protocol for ATM signaling (SPANS). FORE's ForeThought management software runs on Cabletron's Spectrum network management platform so network managers can view ATM-to-Ethernet data flow. Cabletron has developed similar ATM access modules for the MMAC and MicroMMAC.

Features associated with Cabletron switches

Cabletron's SmartSwitch family of switches offer a number of features, including per-port RMON support and embedded network trend and analysis tools. They leverage

features associated with modular hubs, such as fault tolerance, port-level management, thresholding, scalability, and redundant power. All Cabletron switches are
fully manageable by an SNMP-based network management system.

Synthesis

Synthesis is Cabletron Systems' strategic framework for its infrastructure products
and technologies, automated management tools, and support services. The company
also refers to Synthesis as its vision for the future and its plan to migrate router-based
legacy LANs to switched-based networks over time. With tens of thousands of MMAC
hubs installed, Cabletron needs to ensure that its installed base of customers has a
smooth migration path to new technology. The company claims that Synthesis is a
much more cohesive plan than that of its competitors. It features high-speed
switches, layer-3 routing, and automated management software and support services.

According to Cabletron, one of the problems with most of today's virtual LANs is
that they still rely on traditional multiprotocol routers. These routers perform some
necessary functions, such as handling all addressing and policy issues associated
with the decision-making required for forwarding and filtering packets. Under Synthesis, routing functions becomes integrated into the network to create "virtual
routers." The routing functions such as access control and configuration responsibilities are moved into a software-based enterprise management system. Network
managers can limit the traffic from specific areas by setting the policy on a network
device so firewalls can be established wherever they're needed on a network. With
Synthesis, users request a service and a dedicated connection is established by this
distributed management system. Cabletron has promised that the various transport
technologies found today (Ethernet, Token Ring, ATM, SNA, etc.) will all interoperate and act together to form a connection-oriented ATM network.

The infrastructure under Synthesis

Cabletron defines *infrastructure* as the networking devices and technologies that
make up the physical foundation of an enterprise network. Included in this category
are Cabletron's modular and workgroup hubs, routing technology, SNA/LAN integration products, LAN and ATM switches, and the company's virtual network services
(VNS). Cabletron's design principle of adding to its product line without radically
changing the core design is known as its PLUS architecture. One of the key principles
of PLUS architecture is that Cabletron's various platforms (MicroMMAC, MMAC, and
MMAC-Plus) are all scalable because they're based on Intel I960 microprocessors and
C++ programming. Cabletron points out that its platforms began by supporting bridging, later supported routing when this became necessary, and now support packet
switching. The company's two newest infrastructure products are virtual network
services (VNS) and SecureFast switching (SFS), both of which I'll describe shortly.

Automated management under Synthesis

Cabletron believes that network managers historically have used enterprise network
management systems to configure and to monitor devices but did not involve these

systems in real-time operations. The company believes that in the future efficient enterprise network management will require more automation because connection-oriented, switched virtual networks will involve far too many requests for service than can be handled by a human network manager. Under Synthesis, new Spectrum network management applications will be available to work with VNS and automate the management and control of a switched virtual enterprise internetwork. We'll discuss some specifics about Spectrum a bit later in this chapter.

Support services under Synthesis

Cabletron believes that support services under Synthesis will play a major role in helping customers migrate from legacy LAN environments to a switched network environments. The company points out that it's the only network company that now provides 24-hour, seven-day-a-week worldwide coverage with free telephone support during normal business hours. Cabletron is proud of its current record of resolving 98.8 percent of problems on the first call.

Securefast switching

Securefast switching (SFS) is Cabletron's term to describe its virtual network services (VNS), which provide advanced virtual network capabilities for its family of switches to create a switched-based network infrastructure. Under SFS, network managers can specify quality of service. This product family covers both packet- and ATM cell-switching products, all of which share a common set of automated connection management services (ACMS). Cabletron has renamed ACMS *virtual network services (VNS)* technology. Cabletron is quick to point out that SFS can work with or without traditional routers. Today's network environment generally means a switched LAN environment with a router used for wide area network connections. The ability to incorporate traditional routers in SFS means that customers can use existing installed base equipment and protect their investment in this equipment. Figure 9.5 shows the building blocks associated with SecureFast virtual networking.

Spectrum and virtual network management

Cabletron indicates that it will provide a number of virtual network management applications, including policy-based management, configuration management, and network usage accountability. The company intends to provide Spectrum with the ability to allow network administrators to assign users their own access privileges, which will stay with users even if they move from one LAN to another LAN. A graphical user interface that Cabletron describes as intuitive should make it possible for nontechnical people to define usage rules for policy-based management. A second Spectrum application will automate such typically repetitive configuration management tasks as configuring multiple devices, verifying configurations, and scheduling backups during off hours. One of Cabletron's goals is to make it possible for even the most nontechnical people to make additions, moves, and changes on a network. Finally, Cabletron promises that under Spectrum network managers will be able to track and record LAN and WAN bandwidth usage.

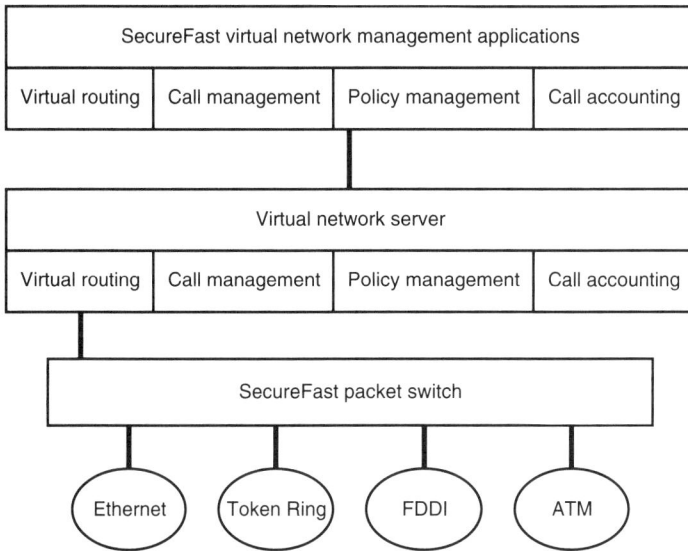

Figure 9.5 The building blocks of SecureFast virtual networking.

Spectrum and management of other vendors' products

Cabletron's Spectrum was originally designed to manage all of its network products. The company has developed a 4.0 version of this product that accepts modules that manage multivendor internetworks that include products from companies such as Cisco, Bay Networks, 3Com, and UB Networks. This product breakthrough gives Cabletron a big edge over its competition because it helps negate the penetration by Bay and Cisco into key accounts. It means that a company that has Cisco or 3Com routers or some Bay Network concentrators can consider Cabletron to manage the entire enterprise network. Spectrum will go far beyond mere generic SNMP management and manage competitors' products at the port level.

Another reason why Spectrum version 4.0 makes Cabletron a much more attractive enterprise network management tool is that it ties together distributed SpectroServer databases. This means that a different SpectroServer can be located at each site within a wide area network to perform alarm filtering at different hours of the day. Results of the alarm filtering are reported back to the central Spectrum management console.

Spectrum also offers another unique feature: Web-based reporting. The program automatically generates reports that can be accessed with any graphical browser, such as Mosaic or Netscape. Because information can be viewed on the World Wide Web, the program eliminates the need to manually produce and distribute network performance reports.

Should Cabletron be your ATM vendor?

Cabletron offers a comprehensive switching network lineup as well as concrete architecture for a smooth migration from legacy LANs to ATM. Perhaps its major long-term strength is the advanced state of Spectrum. Companies that already have other vendors' hubs might seriously consider Cabletron because Spectrum can tie together a multivendor enterprise environment. Because Cabletron generally uses its own salespeople to call on customers and focuses on large companies, it's unlikely that a medium-sized or smaller company would ever have a relationship with a Cabletron salesperson. These companies are much more likely to be contacted and supported by companies such as 3Com and Bay Networks that use value-added resellers and distributors to reach them.

UB Networks

UB Networks has been around a very long time and has a significant worldwide installed base. Its ATM strategy is built around helping its customers protect their equipment investment while offering them a smooth migration path to ATM. Among the partners it has been working with to achieve this goal are BBN, Fujitsu Microelectronics, and Newbridge. The company has developed an "ATM anywhere" strategy. It says that it wants its customers to be able to introduce ATM into a network on an as-needed basis and continue to be able to use their existing Access/One hub modules. A second way the company differentiates itself and its ATM strategy from its competitors is its emphasis on fault tolerance. UB Networks has borrowed fault-tolerant concepts from its parent Tandem Computers to make all major subsystem components redundant. This is important because many of its customers with Access/One hubs support applications 24 hours a day. The third way the company differentiates itself and its products is through scalability of its ATM switches. This scalability will become apparent when you look at some of its switches.

The GeoLAN/500 is an ATM switching hub designed to anchor collapsed backbones by linking switched and shared LANs to one another and to an ATM backbone. Its SuperSwitch architecture, one that comprises Access/One, MultiLAN, and ATM backplanes, lets customers preserve their investment in Access/One hub modules while adopting new GeoLAN/500 cards. Its 17-slot chassis includes 12 slots for Access/One and GeoLAN/500 modules, two center slots for redundant segment switching cards, two slots for redundant ATM switch fabric with a "shortest path" connection for maximum performance, and one dedicated slot for a standard environmental monitoring module.

Scalability is very important to many UB Networks customers. The GeoLAN/500 is scalable to support up to 92 Ethernet, 11 Token Ring, 11 FDDI, or 22 ATM networks. Its SuperSwitch scalable architecture supports one optional single or dual 155-Mbps or 622-Mbps ATM switching fabrics. Its integrated center switch module internally connects all Ethernet segments at wire speed through LAN switching.

UB Networks touts the GeoLAN hub's high-density LAN switching capabilities for creating networks with high-performance switched desktops and/or switched, collapsed workgroups. It points to the switching hub's ability to support up to 92 switched Ethernet workgroups.

The GeoSwitch 155 is an ATM switch with a 5-Gbps nonblocking switching capacity. It can connect up to 16 workstations, hubs, or servers and is supported through SNMP management. This standards-based switch is compliant with ATM Forum 3.0

and 3.1 standards. Any combination of 4, 8, 12, or 16 ATM UNI ports can support 155-Mbps traffic. The switch supports multimode fiber, single-mode fiber, and twisted-pair versions. Integrated onboard processors support embedded signaling, LAN emulation, simple network management protocol (SNMP), private network-to-network interface (NNI), and classical IP over ATM.

UB Networks has positioned the GeoSwitch 155 as ideal for a workgroup switch but scalable as a campus backbone to connect hubs, bridges, and routers. Several of these switches can be connected to each other so the aggregate of the ATM ports can create a backbone with OC-3 (155-Mbps), OC-12 (622-Mbps), or even higher bandwidth.

UB Networks believes that customers are looking for one-stop vendor shopping and would prefer to purchase switches and cards from the same vendor. Its GeoNIC adapter cards provide desktop links to the GeoSwitch. Bus types supported include EISA, PCI, graphical I/O, and VME. The company also offers GeoLink uplink cards that lets users gain ATM connectivity from FDDI and Token Ring networks. It uses a combination of Ethernet-to-ATM translation and user network interface signaling to permit native ATM and legacy LANs to connect to ATM based networks.

The company also offers GeoView, an SNMP-based network management application that it says is optimized for ATM workgroups. It has a graphical user interface and runs with UB Networks' NetDirector or Hewlett-Packard's HP OpenView.

Should UB Networks be your ATM vendor?

UB Networks could be the ideal ATM vendor for you if you fit a certain profile. Obviously, if you've already invested in Access/One hubs, the fact that modules from this hub can be used as part of an ATM migration strategy is very attractive. If ATM for the desktop is a major consideration, then UB Networks offers an attractive one-vendor solution with its hubs and switches. Finally, a company that needs to run several different types of networks including ATM might be attracted to UB Networks. Its hubs support ATM and Ethernet modules concurrently.

Summary

Bay Networks' network architecture and plan to migrate legacy LAN users to a switching network environment is known as BaySIS (Bay Networks switched internetworking services). This architecture covers the backbone, the desktop, and even remote-site access and is built on the three fundamental services a network must provide: transport, operations, and policy services. The philosophy of BaySIS is to switch where you can and route where you must. It has a proprietary extension, known as IPNNI, to handle routing in an ATM environment.

Bay Networks has been slow to offer ATM25 desktop connections and has offered 155-Mbps connections. The heart of any enterprise network for Bay Networks is its SynOptics System 5000 hub. It can manage multiple Ethernet, Token Ring, and FDDI networks. Eventually an ATM backplane will be added to enable this hub to integrate Ethernet and Token Ring to ATM switching modules.

While this company offers BaySis as its comprehensive plan for migration to a switching environment, it has been relatively slow in releasing ATM products. In summary, all the pieces of the ATM puzzle do not as yet fit together to form a real BaySis environment.

3Com has been successful in merging products from its many acquisitions into a coherent switching network environment. High-performance scalable networking (HPSN) is the company's strategy for providing high-speed links for growing networks. Among the promises 3Com makes for HPSN are increased performance and manageability, economical upgrades, and routed ATM. The company has placed its bets on collapsed backbones over distributed backbones.

3Com plans to leverage its expertise in designing and developing application-specific integrated circuits (ASICs) for its HPSN products. Its switches support Ethernet, 10Base-T fast Ethernet, Token Ring, FDDI, and ATM. Much of the company's focus on enterprise networks is on its LinkBuilder MSH multiservices hub. This product can support Ethernet, Token Ring, and FDDI now and will soon support both fast Ethernet and ATM. The company's ATM products as well as its legacy LAN products can be managed via the Transcend network management application.

3Com's NetBuilder II bridge/router also figures heavily in the company's HPSN architecture. The company argues that the choice of most companies today is not between routing and switching and that routing complements switching.

Cabletron had been slow to release ATM products, but its partnership with FORE Systems is already paying dividends as ATM products fill in gaps in the MMAC-Plus, MMAC, and MicroMMAC platforms. Synthesis is the company's strategic framework for its infrastructure products and technologies, automated management tools, and support services. It provides Cabletron's vision for the future and a plan to help customers migrate router-based legacy LANs to switched networks.

Securefast switching (SFS) is Cabletron's term to describe its virtual network services, which are designed to provide advanced virtual network capabilities for its family of switches and create a switched-based network infrastructure. SFS supports quality of service. Cabletron's Spectrum network management application supports the company's own products as well as those from competitors. The advanced nature of Spectrum, including its automation features, appears to give the company a strategic advantage over some of its competitors.

10

Wide Area
Network ATM Vendors

In chapter 4 I described the basic concepts associated with wide area networks. In this chapter I'll discuss several vendors offering specific ATM products for the wide area network. Companies without an installed base of customers with traditional time-division multiplexers, such as Cascade Communications, have been able to move more quickly into ATM technology. Other companies have had to develop migration paths so as not to alienate their current customers. In this chapter I'll examine specific ATM switches as well as the various methods these switches use to handle congestion and other traffic management problems. I'll also describe specific migration strategies to show you how these companies are making migration to ATM as painless as possible.

Stratacom

According to President and CEO Dick Moley, Stratacom has always focused on cell switching. Before the growth of interest in ATM technology, Stratacom was interested mostly in mid- to large-size companies that wanted to build integrated voice and data, private, wide area networks. The company had trouble selling to carriers because this group was more comfortable with time-division multiplexing. Moley has pointed out that, while data networking is what drives the building of private wide area networks, the overlaying of voice traffic is what results in monetary savings. As a result, Moley believes that most of the installed Stratacom switches incorporate voice, synchronous, and video traffic across their bandwidth.

As carriers turned toward frame relay, Stratacom found greater interest in its switches because this group wanted to provide data services as well as voice services in order to move companies away from private networks. Stratacom has found similarities and differences among its two types of customers. Both private users and carriers require the ability to administer and diagnose their networks. Carriers have a very special need for billing requirements, so network management is particularly important to them.

Stratacom believes that, as switched virtual circuits become reality rather than hype and as bandwidth on-demand becomes reality, the distinction between a private network and a public network will become murkier and more difficult to make. Companies such as financial institutions that have the economies of scale as well as the security needs to justify private networks will likely continue along that path.

The IPX

This is Stratacom's low-end WAN switch. It's available in 8-slot, 16-slot, and 32-slot configurations. The switch can package voice as well as data and can compress voice traffic. Its frame relay interface provides an entry-level service into cell-based networks and offers a smooth migration path to broadband ATM services.

General Electric has built a WAN consisting of more than 50 IPX switches. This network supports four traffic types: X.25, bisynchronous traffic, SNA, and LANs. There's also a second WAN that uses Stratacom switches to provide some corporate proprietary applications. Eventually, General Electric would like to merge the two WANs, which now consist of around 80 devices.

IGX

The integrated gigabyte exchange (IGX), 1.2-Gbps, ATM switch comes with a 16- or 32-slot configuration and supports sub-T-1, T-1, T-1, T-3, high-speed serial interface (HSSI), and OC-3. It supports ATM Forum's UNI 3.1 as well as Ethernet, Token Ring, and FDDI traffic via Stratacom's EdgeConnect device. A customer who purchases the 16-slot configuration can add to it by purchasing an additional chassis that runs off the same equipment and a module that connects the two chassis via a ribbon cable.

The IGX is unique at this time because it's the only one-box solution for consolidating existing time-division multiplexing environments that have PBX tielines and SNA. It also contains all software required for management and congestion control. To help customers achieve economies of scaling by overlaying data traffic with voice traffic, the IGX comes with compression and silence suppression of voice over ATM. The company believes that as much as 60 percent of the bandwidth in a typical voice transmission doesn't involve voice information at all. This bandwidth can be saved through compression. The IGX can also compress the actual conversations it transmits with three adaptive differential pulse-code modulation (ADPCM) compression schemes. It can squeeze 64-Kbps voice channels into 32, 24, or even 16 Kbps of bandwidth. Stratacom's switch architecture enables feature modules to be attached to different physical interface cards across a midplane. Figure 10.1 illustrates how this architecture works.

The IGX's major strength is its support for legacy protocols. Stratacom has positioned it as ideal for T-1 multiplexer users because multiplexer vendors don't support ATM and ATM vendors don't provide enough legacy support for their switches. As Figure 10.2 shows, two practical uses of the switch are as an IPX concentrator and as a time-division multiplexer replacement.

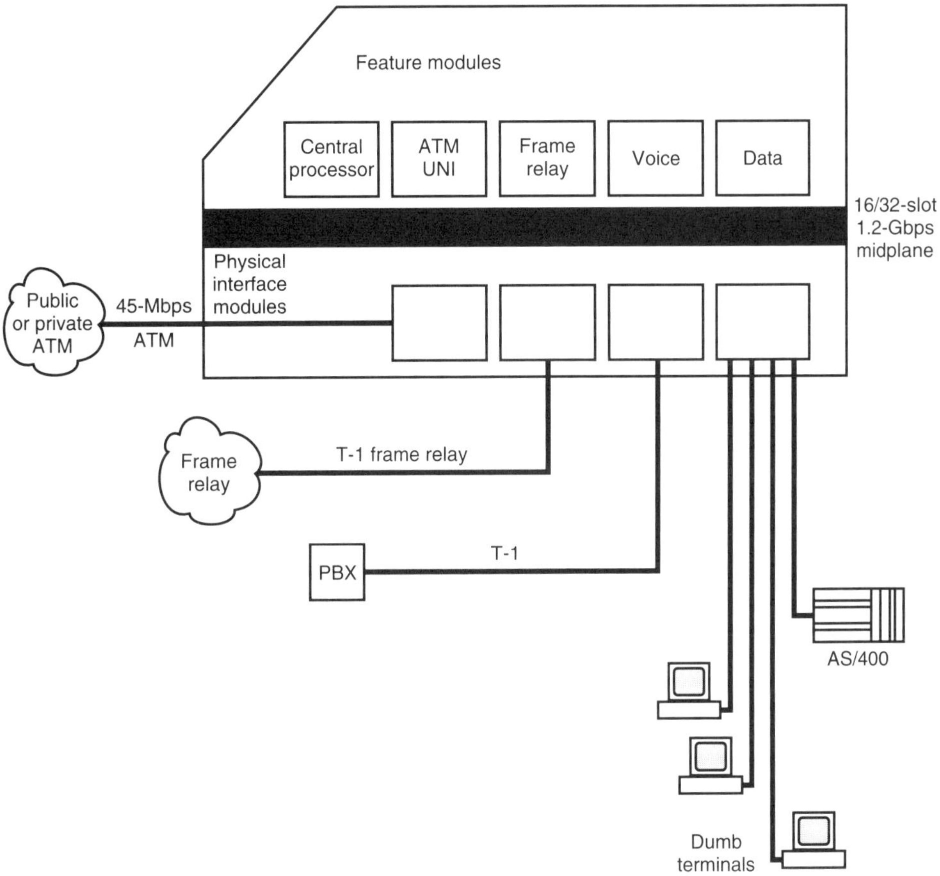

Feature modules

| Central processor | ATM UNI | Frame relay | Voice | Data |

16/32-slot
1.2-Gbps
midplane

Physical
interface
modules

Public or private ATM

45-Mbps

ATM

Frame relay

T-1 frame relay

PBX

T-1

AS/400

Dumb terminals

Figure 10.1 How the Stratcom architecture works.

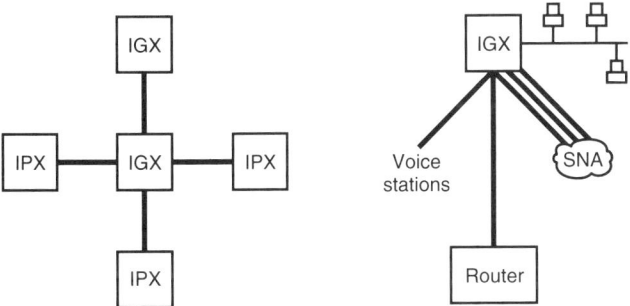

IGX

IPX   IGX   IPX

IPX

IGX

Voice stations

SNA

Router

Figure 10.2 Two uses for the IGX: an IPC concentrator (left) and a
TDM mux replacement (right).

BPX

The BPX is a 10-Gbps ATM backbone switch designed for carriers. The switch offers a number of features, including the following:

- Automatic rerouting if a trunk fails
- ATM-to-frame relay internetworking
- T-1/E-1 circuit emulation, SMDS, HSSI, and OC-3 interfaces
- 32 classes of service

The switch consists of a broadband shelf and up to 16 optional narrowband shelves. The broadband shelf is a 15-slot chassis with a total capacity of OC-12 bandwidth.

AXIS

AXIS is a network access component designed to aggregate user traffic for transmission over broadband ATM networks. It supports several narrowband or slower-speed WAN interfaces, including frame relay, ISDN, switched multimegabit data services (SMDS), ATM, and circuit emulation for the BPX. It enables companies to add up to 16 AXIS units that as an aggregate can provide support for up to 15,000 frame-relay user connections operating at 64 Kbps.

Stratacom software

The software available with the IGX includes ForeSight for closed-loop congestion control, OptiClass for 32 class-of-service definitions, AutoRoute for virtual-circuit management and automatic rerouting, FairShare for per-virtual circuit queuing and rate scheduling, and StrataSphere for ATM management.

OptiClass. OptiClass provides up to 32 programmable network-wide service subclasses. These service subclasses can be further defined by the user and assigned to connections. This software provides a minimum bandwidth guarantee. The *minimum cell rate (MCR)* provides deterministic performance per connection and per subclass so one type of traffic isn't frozen out by another type of traffic. There are also per-VC queues and per-VC rate scheduling firewall connections to prevent misbehavior of one class from affecting another class. Figure 10.3 illustrates the functions associated with OptiClass.

AutoRoute. AutoRoute connection management software automates management of virtual circuits and bandwidth allocation by routing and rerouting virtual connections over optimal paths through the network. It can track resources designated to individual connections to prevent system overloading and ensure users of their committed information rate (CIR) at all times. If there's a trunk or switch failure, AutoRoute will reroute virtual circuits to alternate network paths.

ForeSight. ForeSight provides congestion avoidance as a closed-loop, rate-based feedback mechanism that looks ahead through the network to determine whether

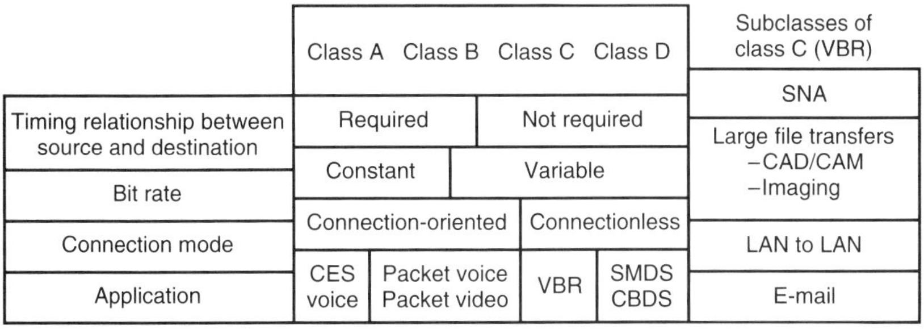

	Class A	Class B	Class C	Class D	Subclasses of class C (VBR)
					SNA
Timing relationship between source and destination	Required		Not required		Large file transfers −CAD/CAM −Imaging
Bit rate	Constant		Variable		
Connection mode	Connection-oriented			Connectionless	LAN to LAN
Application	CES voice	Packet voice Packet video	VBR	SMDS CBDS	E-mail

Figure 10.3 The functions associated with OptiClass.

spare capacity exists. It can allocate spare capacity among active virtual circuits to optimize resource usage and avoid congestion. ForeSight enables Stratacom's switches to support available bit rate (ABR) services. Stratacom argues that ForeSight permits up to 95% usage of network bandwidth, compared to from 50% to 70% usage with open-loop mechanisms. According to Stratacom, the benefits of closed-loop feedback mechanisms are related to their use of separate queues for each ATM virtual connection. Connection-based queuing establishes firewalls among network users to ensure predictable quality of service regardless of operating conditions. ForeSight uses FairShare per-VC queuing and per-VC rate scheduling, as well as rapid cell-based feedback to prevent congestion.

StrataSphere network management

StrataSphere provides fault isolation, performance, and configuration management of Stratus's ATM networks. It collects statistics that can be filtered in predefined formats for performance and billing purposes. StrataSphere BILLder is an application under StrataSphere that monitors a network's traffic flow within a user-defined billing period and formats the data from frame-relay and ATM connection endpoints into a standard Bellcore automatic message accounting (AMA) or customized billing records.

StrataSphere connection manager uses a forms-driven interface to provide integrated provisioning for connections based on quality-of-service parameters such as the type of connection being made and the available bandwidth. The service connections are established using SNMP and will also support the OSI's common management information Protocol (CMIP) in the future.

Other management tools include Modeler, Optimizer, and StrataView Plus. StrataSphere Modeler provides synthesized network modeling using existing network configuration data. StrataSphere Optimizer enables customers to perform "what if" scenarios based on predefined parameters. Finally, StrataView Plus is an SNMP-based network management platform available under Hewlett Packard's OpenView as well as NetView for AIZ, which provides fault, configuration, and performance management. Gathered information can be stored in an Informix SQL database for integration into customers' existing multivendor systems.

Strategic partnerships

Stratacom has developed several partnerships and alliances with major industry players. It has collaborated with AT&T and Cisco on frame-relay technology and ATM internetworking. Its agreement with Bay Networks calls for development of a migration path for LAN-to-WAN SVC-based ATM networking that incorporates Bay Network's LattisCell ATM and Stratacom's BPX ATM switches. From a network management perspective, the two companies want to integrate Bay's Optivity software with Stratacom's StratView. The two companies have also agreed that they want to integrate the same flow control and class-of-service methods in their products. Finally, the companies will develop frame relay and T-1 ATM interfaces for the LattisCell switch for LAN-to-WAN communications without the need for routers.

Stratacom also has an agreement with NetEdge to integrate Stratacom's switching and NetEdge's knowledge of edge routing. This agreement covers developing StratView's ability to manage EdgeConnect products and ensures both that there's interoperability between the two companies' SVC signaling and that there's frame-relay compatibility between the BPX switch and the EdgeConnect router.

Newbridge Networks

Newbridge Networks stands alone with IBM as the only vendors who offer both wide area network and local area network ATM switches. It has divided its attention between T-1 multiplexers, frame-relay packet switches, and ATM switches, and is the leader in T-1 time-division multiplexers. Of the approximate $100 million it spends on research and development, $30 million is spent on ATM research. One of the company's major strengths is that network management software spans its entire spectrum of products. In 1993, the company shipped its 36150 MainStreet ATMnet, a wide area ATM switch. It also launched VIVID development, a LAN ATM family of products.

The 36150 MainStreet ATM network switch

The 36150 MainStreet switch is a wide area network ATM edge switch. While it can be used as a core backbone switch in smaller networks, it's optimized for use as an edge switch with features such as native LAN and video interfaces. The 36170 MainStreet ATMnet is a backbone switch designed for service providers and large corporate users.

Cards support four-port Ethernet, three-port Token Ring, and a single-port, 100-Mbps, FDDI interface. There's also support for a four-port T-1 ATM interface and a T-3 interface. This switch and the 36170 represent Newbridge's attempt to bring its first-generation MainStreet switch family up to a position where it's competitive. The 36170 supports frame relay as well as network and service internetworking between frame relay and ATM. It has a switching capacity of 12.8 Gbps with planned expansion to 102.4 Gbps.

The VIVID workgroup switch

The VIVID workgroup switch is designed for LAN backbones and workgroups and supports routed LAN emulation. It has a 1.6-Gbps nonblocking switching capacity and twelve 155-Mbps ATM interfaces. It supports switched virtual circuits, the ATM Forum's UNI, and SNMP network management. This switch is designed to connect VIVID ridges (described in a later section), ATM workstations, and hubs and routers equipped with ATM interfaces.

Traffic management

Each of the VIVID workgroup switch's output port is equipped with four variable-size buffers that can be programmed to support multiple qualities of service. The switch services the buffers on an exhaustive priority basis to ensure that applications with high priority aren't blocked by low-priority application traffic.

VIVID network interface cards

The VIVID product line currently supports EISA, GIO, and bus platforms for ATM network interface cards. These cards receive a stream of ATM cells, demultiplex them into their respective virtual connections, and then reassembles the discrete cell streams into the native data structures. The cards perform packet-level error checking as specified by the appropriate ATM adaption layer (AAL).

Newbridge's VIVID solution for legacy LANs

There are several components to a VIVID network that incorporates both ATM and legacy LANs. These components support the ATM Forum's specifications for LAN emulation. The VIVID ATM hub is a nonblocking ATM switching matrix that can be configured with 4 to 32 ATM switch ports. The bandwidth ranges from 640 Mbps to 10 Gbps, and the hub supports ATM, Ethernet, Token Ring, FDDI, and video NTSC/JPEG. The ATM interfaces supported include the following:

- ATM (100 and 140 Mbps)

- OC3 ATM (155 Mbps)

- T3 ATM (45 Mbps)

The hub will soon support a 622-Mbps ATM transmission speed. Newbridge argues that it's too expensive to add an ATM interface for each PC and workstation, so it has created VIVID LAN service units (LSUs) and VIVID ridge products to concentrate legacy LAN traffic to VIVID ATM hubs.

The VIVID LSU concentrates Ethernet traffic to forward it to a VIVID ATM hub. It can be configured with 12 or 24 Ethernet 10BaseT ports. It comes with a backbone Ethernet port (10 Mbps) for connection to a VIVID ATM hub. Support is also provided for 10BaseT, 10Base2, AUI, and FOIRL.

The VIVID ridge family of products serves as an interface between shared-media LANs and an ATM backbone. They can handle packet forwarding and cell conversion. The VIVID yellow ridge is designed to provide wire-speed switching among all Ethernet and ATM ridge ports. It comes with 12 Ethernet ports and 1 or 2 LATM ports (100 or 140 Mbps) for connection to a VIVID ATM hub. It supports local segmentation and reassembly (SAR) of Ethernet packets into ATM cells. The yellow ridge performs standard IEEE 802.1d transparent bridging as well as routing at 178,560 packets per second.

The VIVID blue ridge is designed to support Token Ring LANs. It comes with 8 Token Ring ports (4 or 16 Mbps) and 1 or 2 LATM ports (100 or 140 Mbps) for connection to a VIVID ATM hub. It also supports local segmentation and assembly (AR) of Token Ring packets into ATM cells.

VIVID architecture unbundles traditional router functionality by separating the forwarding functions from the routing servers. Forwarding functions are performed by ridges, NICs, and ATM switches, while the routing services are performed by the route server. The VIVID route server functions as a virtual router control card. It performs all route calculation and distributes routing table updates to ridges and other network components as required. It also processes all broadcast traffic, such as ARP requests. Rather than flooding these messages to all attached devices, ridges forward them directly to the route server, which processes the broadcast message and directs the response back to the appropriate node.

The VIVID system manager monitors, controls, and troubleshoots a network and facilitates virtual connections through the ATM switch network. It monitors components, provides ATM-based connection management, and configures virtual LANs. The system manager offers a graphical user interface to display maps and device drawings. Users can point and click to create, populate, and assign filters to virtual subnets. Only one virtual router needs to be configured. It in turn distributes routing information to the proper forwarding elements.

The system manager offers automatic device and topology discovery, logical and physical network views, and fault detection. It also offers performance monitoring and statistics as well as in-band and out-of-band management.

Figure 10.4 shows the VIVID product family linking together corporate network resources. It also demonstrates the role of Newbridge's wide area network 36150 MainStreet ATMnet.

Newbridge and virtual networks

Newbridge believes that its focus on routing helps to differentiate it from other vendors. VIVID provides virtual subnetworks. Traffic within a virtual subnet is forwarded the same way local traffic on a segment connected to a router port would be, and traffic between virtual subnets is routed using network layer protocols that adhere to firewall restrictions. Newbridge's virtual networks also enhance security because VIVID architecture provides the option of secure ports. ATM or ridge ports associated with a particular device address or virtual subnet can be designated as secure ports. If an unauthorized device address appears on the secure port, that device is denied access to the network (no traffic is allowed to flow) and appropriate management reports are generated.

VIVID and SMC

Newbridge's VIVID business unit and Standard Microsystems Corporation (SMC) have a strategic alliance that includes technology sharing and codevelopment. The

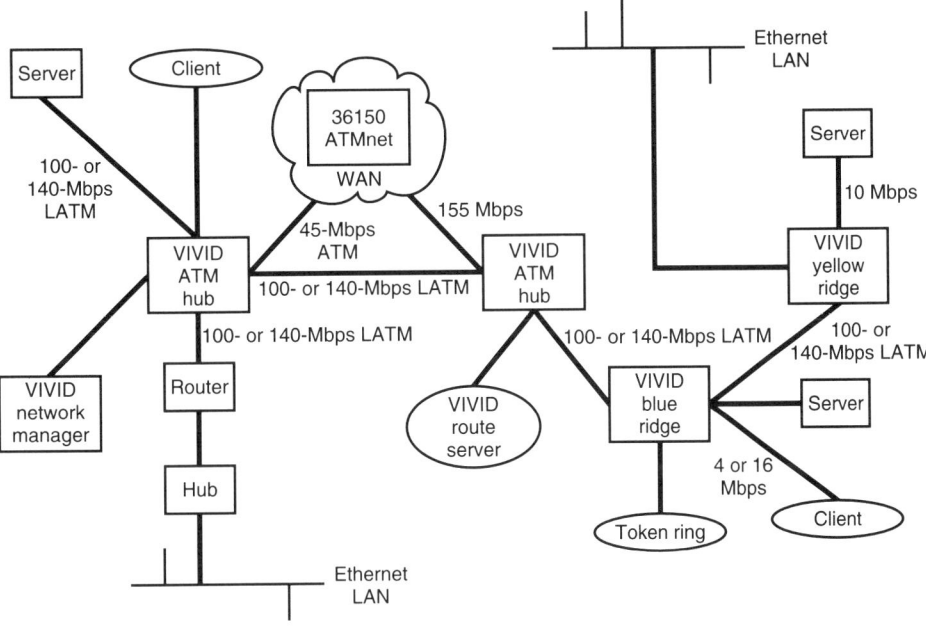

Figure 10.4 The VIVID product family in action.

goal is to achieve seamless integration of LAN and WAN ATM products. The alliance is beneficial to both companies. SMC is able to tell customers that their LAN ATM will be interoperable over Newbridge's WAN ATM links. For Newbridge, the major benefit is access to SMC's competency in designing and manufacturing ATM silicon, bus interfaces, and LAN internetworking devices. These competencies will be used by Newbridge in developing enterprise networking products. Table 10.1 shows how the two companies envision their roles in this alliance.

The two companies will develop low-cost workgroup ATM switches (12 to 16 ports) and will provide ATM support for EISA, ISA, PCI, and Sbus systems. From a practical perspective, the companies sales channels are very complementary. Newbridge sells direct while SMC uses volume distribution channels.

TABLE 10.1 The Newbridge/SMC Alliance

Product	SMC's role	Newbridge's role
Campus ATM switch	OEMing and integrating technology into future products	Contributing core technology
Workgroup ATM switch	Joint development	Joint development
Adapter cards	Contributing core technology and leading product development	OEMing and integrating technology into other Newbridge products

Some Newbridge strengths and weaknesses

Newbridge antagonized its customers by promising ATM products and then failing to meet deadlines. It has a coherent, well-thought-out ATM architecture (VIVID) and a unit devoted just to that part of its operations. It has been criticized by some customers who characterize the MainStreet switch as still being a first-generation product. More serious is the price points of its switches, which aren't currently competitive. The company faces increasing competition in the LAN area and will have to leverage its alliance with SMC to offer a complete ATM solution. Because SMC sells through distributors and VARs, one problem for Newbridge's direct sales force will be to locate the prospects for enterprise Newbridge ATM products that already have SMC products installed.

Cascade Communications

Cascade focuses entirely on ATM for data, and the company believes that it offers the widest array of data trunking options of any switch on the market. Cascade describes the carrier market as two-tiered, with its emphasis on the second tier. The first tier represents a network consisting of high-throughput ATM switching fabrics interconnected with SONET transmission facilities. These switching fabrics scale in multiples of tens of gigabits per second and also handle bulk transmission facilities in the DS-3 to OC-48 range. The large, expensive switches in this tier are found in central offices and are the core of the carriers' network.

Cascade has chosen to concentrate on the second tier, which it describes as one of emerging carrier networks. This tier is located closer to customers physically and is based on multiservice WAN switches, sometimes described as *access switches* or *edge switches*. Functioning as the customer's access point, these switches place a premium on service and flexibility. Carriers use such switches to offer services such as frame relay, SMDS, and ATM, as well as open systems interconnection (OSI) and internet protocol (IP) routing over T-1 or T-3 links. These switches also provide the internetworking capability to link frame relay and SMDS to ATM.

Switches

Cascade's family of STDX and B-STDX switches are positioned as multiservice WAN switches to perform in a public network as edge switches or second-tier switches that extend services from central office switches to points closer to customers. The B-STDX 9000 extends the STDX family to support ATM as well as frame relay and SMDS. It has a 16-slot modular ATM internetworking platform and can be configured with redundant control processors, intelligent I/O modules, fans, and power supplies. Among the I/O modules planned for this switch are the following:

- Dual-port, high-speed, serial-interface (HSSI) module permitting network access up to 45 Mbps

- ATM I/O module providing a standard ATM DXI interface to an ATM public network

- A 10-port DSX-1 for high port density

Because Cascade uses a symmetrical architecture, all ports are software-defined as trunks, network-to-network interface (NNI) links, or user-to-network interface (UNI) links. Any port can be used for any of the software-configurable interfaces, whether it's frame relay, SMDS, or ATM services. For example, in a configuration where branch offices connect via frame relay to the B-STDX 9000, the switch can then trunk to the ATM backbone at T-3 speeds to ensure high-performance services in the ATM wide area network infrastructure. MFS Datanet has chosen the B-STDX 9000 switch for its frame transport service to support frame relay over an ATM network. Figure 10.5 shows the Cascade B-STDX 9000 in an ATM environment.

The Cascade 500 matrix switch is positioned as a backbone switch for ATM wide area networks. It supports 56 OC-3 ports, 112 DS-3 ports, and 14 OC-12 ports for high-speed lines ranging from 1.55 Mbps to 622 Mbps. There is an aggregate non-blocking capacity of 2.5 Gbps or 5 Gbps, expandable to 10 Gbps. This switch takes a different approach toward providing quality of service for individual connections. Rather than placing the function in software, Cascade gives this responsibility to the switch's backplane.

The Cascade 500 offers four parallel switch fabrics to provide separate cell buffers for each of the four classes of service: constant bit rate (CBR) for traditional voice and video circuit traffic on an ATM backbone, variable real-time bit rate (VBR-RT) for interactive video sent via compression, variable non-real-time bit rate (VBR-NRT) for video broadcasts where transmission delay isn't an issue, and available un-specified bit rate (ABR/UB) for bursty traffic.

Cells are switched into one of the four parallel dedicated buffer structures using an internal switch-routing header to recognize cells that should be sent to a particu-

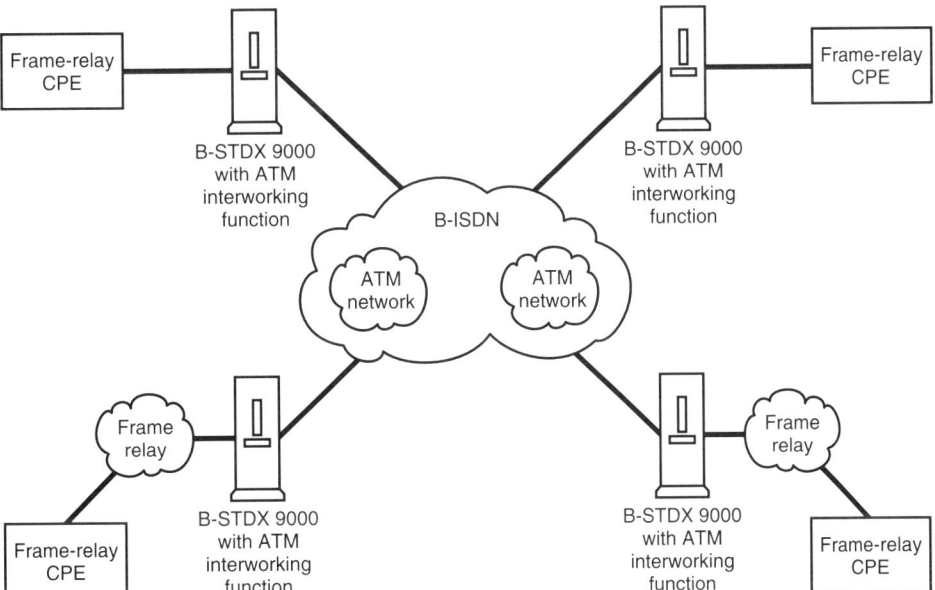

Figure 10.5 Cascade's B-STDX 9000 in an ATM environment.

lar backplane. The buffers supporting VBR traffic are further subdivided with four individual threshold markers, for a total of 10 quality-of-service classes. The ATM switch fabric has a total of 128,000 cell buffers partitioned into four programmable planes. In addition, input/output (I/O) buffers offer additional levels of buffering for each module. The switch offers a dedicated switched virtual-circuit (SVC) signaling processor on every line card for signaling performance scalable according to port density.

Cascade thinks that the combination of the 500 and the B-STDX is particularly powerful. While the B-STDX has a lot of frame links, it's limited to a 1.2-Gbps backplane. Frame-relay traffic can come into the B-STDX and be converted to ATM. It can then go to the 500 to be switched.

OPTimum software

Open packet trunking (OPTimum) software enables public packet networks such as frame relay, ATM, and SMDS to be used as trunk links between switches rather than expensive leased lines. Customers can mix and match public packet services, such as using low-cost shared packet networks in combination with private networks.

Network management

As indicated earlier, Cascade switches support simple network management protocol (SNMP). The company offers the CascadeView/UX network management station (NMS), which supports a Cascade frame relay, SMDS, and ATM switch network. This system provides centralized management, including network configuration, performance monitoring, fault management, and security for all switches on a network via a single connection to any one of the switches.

Strategic alliances

Cascade Communications has a joint technology and marketing relationship with Cisco Systems. They have agreed to jointly develop a multiservice WAN solution for the telecommunications marketplace based on combined Cisco and Cascade technology.

The two companies have agreed to market Cascade's products together. Furthermore, under the agreement Cisco licenses the internet protocol (IP) and open systems interconnect (OSI), routing technology modules of its internetwork operating system (IOS) software, to Cascade for use in product development so that Cascade's products will have seamless interoperability with those of Cisco.

Because of Cisco Systems' position as the dominant leader in the routing marketplace, the agreement helps Cascade convince its customers that its switches will be interoperable with any Cisco routers.

General Datacomm

General Datacomm has become a major ATM vendor while working hard to preserve the significant investment its installed base has made in traditional time-division multiplexers. The company has identified the following four classes of ATM implementation:

Class 1. This class of switches serves as the hub for workgroups and LANs, and also provides local switching and multiprotocol access of up to 2.5 Gbps to ATM backbone networks.

Class 2. This class of switches serve as the enterprise ATM switches for private networks and have throughputs ranging from 2.5 Gbps to 10 Gbps.

Class 3. The third class of switches is designed for service providers, and acts as the public network feeder with throughputs ranging from 2.5 Gbps to 10 Gbps.

Class 4. This class of switches serves in the central offices of service providers, and offers throughputs exceeding 20 Gbps.

The company has stated that it's a key player in class 1 through industry partnerships, and class 2 and class 3 through contract wins.

Strategic advantages

The APEX-Advantage is the first comprehensive ATM network architecture to support simultaneous data, voice, and video applications. The architecture allows switches to provide virtual-circuit management by employing distributed connection management as well as semipermanent virtual circuits that ensure both network availability and network scalability. GDC prefers customers to view data-only ATM switches as being ideal for workgroups and backbones at the edges of an enterprise network. The company admits that as of yet there are no compelling reason for data, voice, and video to share wires within a building, and that data-only switches can keep the per-port cost down.

Having conceded these points, however, GDC is quick to argue that there are several exciting new applications, such as distance learning, that require high-quality video as well as audio to accompany data. And there are several other applications, such as teleradiology and collaborative computing, that are more valuable if transmitted over a wide area network. Also, because telephone calls account for the largest portion of most companies' communications bills, it would make sense to purchase an ATM switch that could handle voice communications efficiently enough to pay for the data and video traffic. Furthermore, GDC stresses that an effective ATM network architecture must make use of narrowband connections as well as existing time-division multiplexers, PBXs, video codecs, and legacy data devices in order to protect equipment investments.

Each interface module in an APEX-ATM switch includes a connection management processor that interacts with its peers throughout the network to create ATM virtual circuits that comprise both virtual channels and virtual paths. GDC argues that this approach is beneficial because, as interfaces are added to an APEX network, the number of connection managers increases to ensure that call setup capacity grows with the network. Every time a new PVC is created, the distribution connection managers fulfill the request by creating internal SVCs along the selected path. If an outage occurs within the APEX network, new SVCs are quickly created along an alternate path so the user's virtual circuit isn't disrupted. Figure 10.6 illustrates the value of semipermanent virtual circuits.

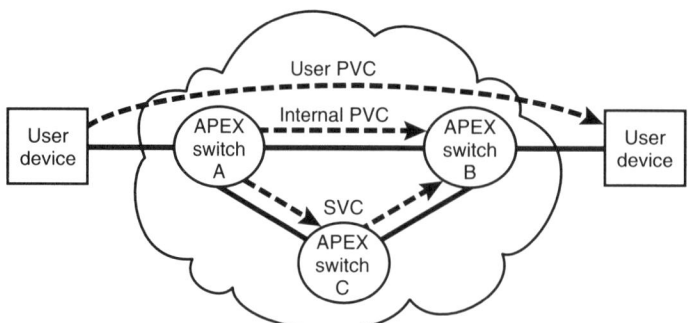

Figure 10.6 The value of semipermanent virtual circuits.

Because the internal SVCs are created with the standard Q.2931 signaling be-tween APEX-ATM switches, a public ATM network can be inserted between APEX switches and automatically participate in the creation of internal SVCs.

ATM switches

GDC's scalable family of ATM switches covers the gamut of campus, wide area, and access switching, as well as hybrid public/private ATM networks. The APEX switch family includes the DV2 corporate network switch, the NPX carrier provisioning switch, and the APEX-MAC, a low-end multimedia access concentrator. The DV2 is designed for backbone transport in corporate LAN-based networks. It can handle HDLC and SNA/SDLC traffic, switched frame-relay traffic, and circuit-switched data, voice, and video by using circuit emulation. It contains Ethernet interfaces as well as ATM interfaces, and includes 19 slots, two of which are reserved for the switch fab-ric. A 3.2-Gbps throughput and 6.4-Gbps switch fabric are available. All active com-ponents are configured redundantly for ensure uninterrupted operation.

The APEX-NPX (network provisioning exchange) switch is similar to the DV2 but is positioned as a low-cost switching platform for carriers and cable TV operators. It has a capacity of 32 SONET/SDH interfaces, supports all ATM standards including SVC (Q.2931), and contains advanced traffic-management functions including policing.

While the APEX-MAC and APEX-MAC1, with 8 and 15 slots respectively, are smaller configurations that are positioned as smaller local backbones. They're also positioned as edge switches that can concentrate access from remote sites. Finally, they could also compete in the ATM workgroup switch market.

All modules in the APEX are compatible so, if a site outgrows one switch, the mod-ules purchased for that switch can be used in another APEX switch. All switches support an SNMP-based network management system. They also support switched virtual circuits as well as permanent circuits. GDC provides software support for congestion management and traffic prioritizing on its 6.4-Gbps switching fabric.

Video support

GDC is planning several video interfaces for its ATM switches, including motion JPEG-based broadband video, H.320 video conferencing, MPEG-1 and MPEG-2 digital video standards, and broadcast-quality video. A 128-user multimedia multipoint server developed by GDC also allows a network administrator to add video and audio support over an ATM LAN.

Traffic management

APEX usage parameter control (UPC) follows ATM Forum UNI specifications and provides four "leaky buckets" for each virtual circuit in order to control traffic entering the network. With this scheme it's possible to manage both peak and sustained traffic for cell loss. A distributed buffering scheme enables GDC to provide maximum switch throughput and minimum latency and cell delay variation. Cells that are received pass through the UPC mechanism and are assigned to high- and low-priority input buffers on each I/O module according to the class of services requested. Intermediate buffers prevent any head-of-line blocking within the switch-matrix fabric. Upon leaving the matrix fabric, the cells are assigned to high- and low-priority output buffers, from which they move on to outgoing interfaces. Both constant bit rate (CBR) and variable bit rate (VBR) traffic can be carried with the required quality of service. GDC has indicated that it will support the ATM Forum's standard for available bit rate (ABR) service.

Network management

The APEX-Advantage architecture supports SNMP management protocols and standard platforms such as OpenView. Among the management functions supported are the following:

- Fault management
- Security management
- Trouble tickets
- Diagnostics
- Statistics collection
- Accounting

Strategic partnerships

General Datacomm and Digital Equipment Corporation have a sales, marketing, and technical agreement to offer customers integrated ATM for LANs and WANs. It appears to be an ideal solution for both companies. Digital needs to be able to tell customers at enterprise network sites that it can provide seamless LAN-to-WAN integration. General DataComm has no real LAN presence and needs Digital to establish credibility with data-centric network managers, who aren't familiar with telecommunications vendors. The companies have announced that they will provide voice, video, and data for campuses as well as campus-to-campus and campus-to-remote locations. They will support multiple WAN services, including frame relay, T-1/E-1, T-3/E-3, and ATM. GDC's APEX family already supports both forms of frame relay/ATM internetworking as specified by the ATM Forum and the Frame Relay Forum: frame-relay ATM internetworking and frame-relay ATM service.

Plans for future integration between the two companies' product lines include linking GDC's APEX-ATM product family with Digital's DEChub family, as well as its GigaSwitch/ATM switching system and ATMworks family of network interface adapters. GDC has also been given a royalty-free license to use Digital's congestion-avoidance and flow-control technology.

General Datacomm has an agreement with DSC Communications Corporation for the evolution and integration of GDC's ATM technology into DSC's global telecommunications products. The agreement calls for the two companies to explore opportunities that would combine DSC's networking software and ATM expertise with GDC's APEX-ATM products. DSC will provide product and service coverage to the telecommunications market using GDC's ATM access technology to enhance DSC's core networking capabilities. DSC plans to introduce ATMs products that will complement DSC's IMPAX 5400 series.

GDC also has a pact with NetEdge to sell that company's ATM Connect edge router in conjunction with its APEX-ATM switch. The ATM Connect product is targeted at service providers who plan to offer native LAN data services over ATM networks. Among GDC's major U.S. accounts are MCI and U.S. West.

Migrating customers to ATM

GDC's strategy with customers who have substantial investments in time-division multiplexers is to help them leverage their equipment and still take advantage of ATM where needed. Figure 10.7 illustrates a company's network environment where ATM has replaced an FDDI distributed backbone for crucial LAN-based applications. Application servers are directly connected to ATM switches at SONET/SDH speeds while ATM emulation makes the ATM upgrade transparent to the legacy LAN equipment. From GDC's perspective, the company's next migration step could include linking more voice and video devices directly to the APEX-ATM switches in order to take advantage of the switches' ability to provide integrated voice and video traffic.

ADC Kentrox

ADC Kentrox is another WAN company trying to transition from the T-1/T-3 multiplexer business to the ATM marketplace and provide ATM for its installed base of customers. Its T-1/E-1 ATM access Concentrator (AAC-1) supports voice, data, and video traffic with variable bit rate (VBR) and constant bit rate (CBR) modules available. The switch can accept voice traffic and prioritize its ATM transport to ensure end-to-end delivery. It uses voice-compression technology to minimize the bandwidth needed for CBR traffic.

The AAC-1 is a multiplexer that converts voice, video, and data traffic into ATM cells for transmission over the wide area at T-1/E-1 speeds. The CBR module takes serial video data streams from a V.35 or RS-449 interface of a codec and converts the video for transport over ATM using the adaption layer 1 (AAL1).

The AAC-3 provides support of ATM, frame relay, SMDS, and CBR, so it can function effectively in an hybrid network environment. Figure 10.8 shows this situation. As Figure 10.9 demonstrates, the AAC-3 also complements ATM switches as an eco-

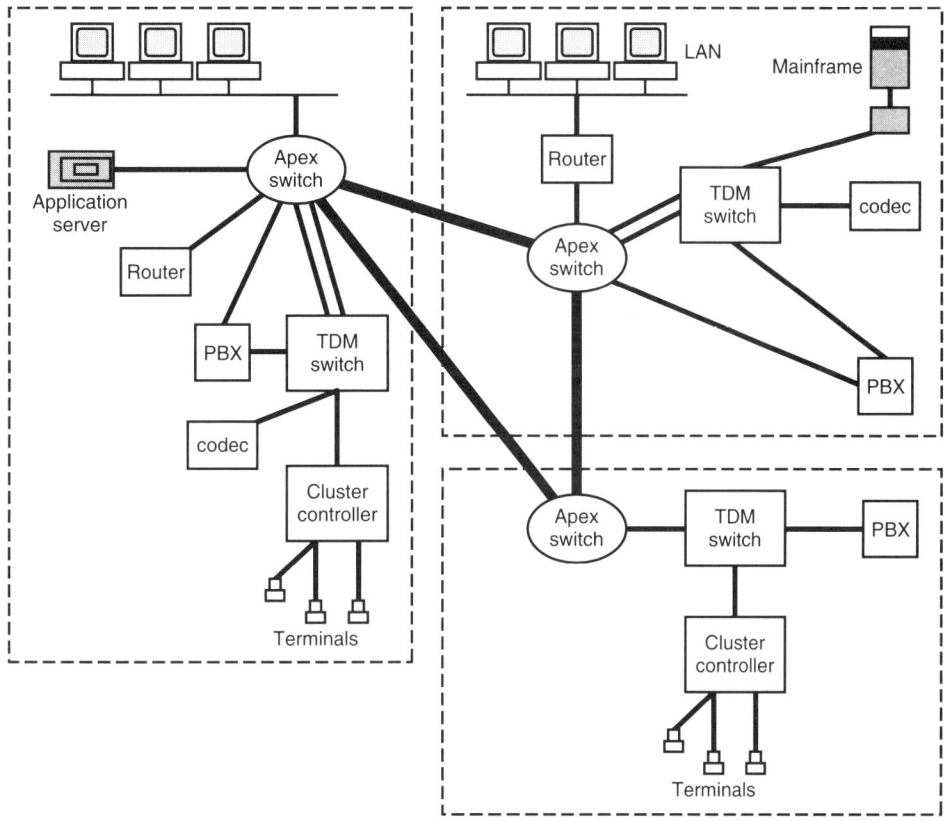

Figure 10.7 A company partially migrates to GDC's ATM.

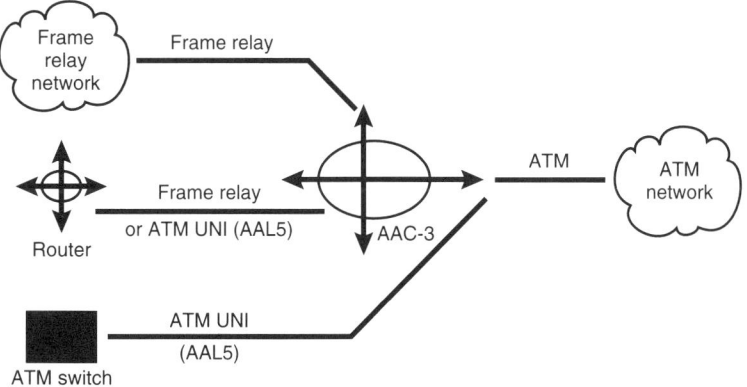

Figure 10.8 The AAC-3 in a hybrid network environment.

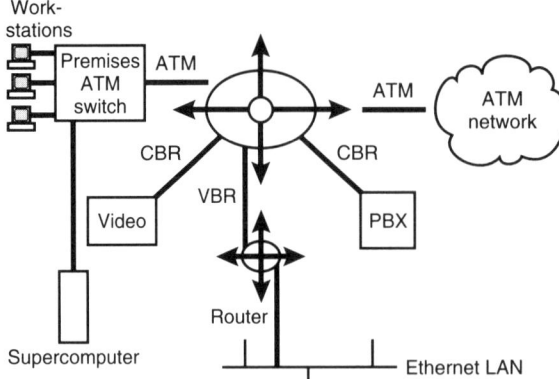

Figure 10.9 The AAC-3 as a fan-out for ATM premise switches.

nomical port fan-out device and access concentrator because it provides several interfaces not available on most ATM switches. Many private networking ATM switches lack the required synchronization to carry time-sensitive CBR traffic. When the AAC-3 is placed in front of these switches, it provides the necessary synchronization as well as the complement of CBR interfaces.

The company has announced interoperability and conformance of its switch with the WilTel ATM network. This means that network managers can work with their carriers to build scalable T-1, T-3, and OC-3 ATM networks.

Northern Telecom (Nortel)

Northern Telecom was a founding member of the ATM Forum and has been actively involved in this technology from the beginning. The Magellan family of enterprise switches includes the DPN-100 product line as well as the Magellan Concorde, Vector, and Passport. The Concorde is Nortel's ATM backbone switch, with a modular switching fabric that ranges from 10 Gbps to 80 Gbps. The Vector is a carrier-grade switch designed for public ATM service. I'll focus on the smallest of these ATM switches, the Passport, which is designed for ATM inter-LAN switching.

The Passport is a single-shelf, 1.6-Gbps switch that can support ATM with the insertion of ATM interface cards. The shelf holds 16 RISC-based interface cards. The switch supports both frame switching and cell switching, and includes a multiple priority system for multimedia communications. Nortel emphasizes that its architecture was designed specifically to handle both frame and cell switching, thus the name FrameCell. Both variable bit rate and constant bit rate traffic is supported. The switch offers 99.999% availability as a result of such features as hardware redundancy, a dual load-sharing bus, trunk backup capabilities, redundant power and processor configurations, and fault-tolerant software. Figure 10.10 shows how Nortel visualizes the role of its Magellan Passport in an ATM environment.

Multiple priority system (MPS)

Northern Telecom's multiple priority system (MPS) allows multiple priorities of traffic in and out of the Passport switch. Each Passport card contains a frame buffer for packet forwarding and queuing. The switch supports traffic management on a link basis to guarantee link bandwidth for constant bit rate traffic. Any bandwidth not used for constant bit rate traffic is available for variable bit rate data traffic. To prevent congestion, the switch provides load sharing across multiple paths, traffic sharing, and buffer allocation. It can react to congestion by rerouting traffic, adapting the rate of traffic at network entry, explicit congestion notification, and being able to prioritize discards.

How Passport will evolve

Northern Telecom has a number of plans to upgrade ATM features in Passport in the near future, including the following:

- Interfaces to support ATM UNI: T-3/E-3, single-mode OC-3, E-2, T-1/E-1

- 155-Mbps trunking for Passport and multimedia services over ATM

- ATM gateway routing supporting termination of ATM UNI (PVC and SVC) on Passport multiprotocol routing (IP, etc. on ATM stacks using AAL5) for internetworking with legacy LANs

- End-to-end ATM networking to provide switching and routing for end-to-end ATM VCs across Passport networks (both PVCs and SVCs supporting variable and continuous bit rate traffic are in the plans)

- Support for multivendor networking by implementing ATM UNI standards and IETF standards

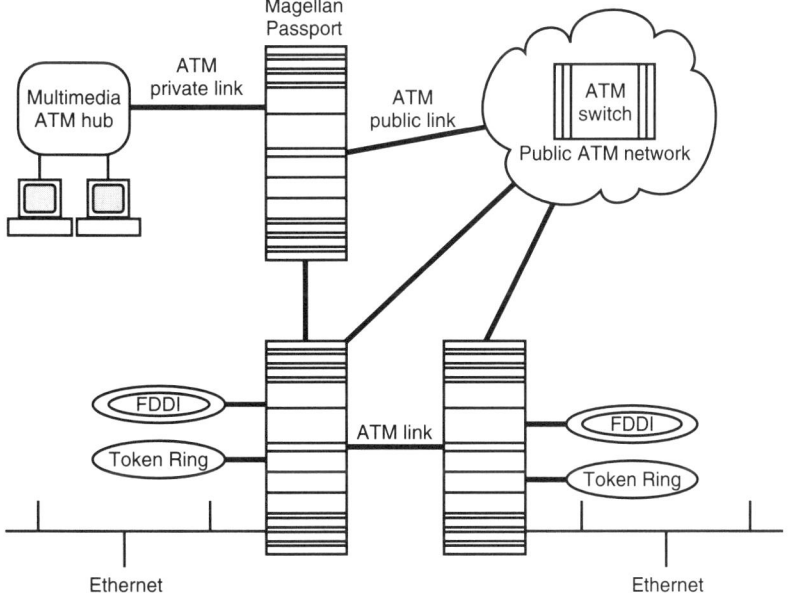

Figure 10.10 Magellan Passport in an ATM environment.

AT&T

AT&T's strategy is to position its GlobeView 2000 broadband system (GCNS) as the perfect product for service providers who want to meet the needs of today's data-intensive customer while still providing a graceful evolution to higher-speed networks. The company's solution includes the following equipment:

- The GlobeView 2000 system service node
- The GlobeView 2000 service management module
- The GlobeView 2000 access module
- The BNS-2000 data server
- The switched services module
- The LCS family of routers and bridges

The service node

The GlobeView 2000 system service node consists of an ATM switching fabric, numerous interfaces, and a fault-tolerant controller. The switch has direct SDH optical interfaces at STM-1 and STM04, as well as SONET rates of STS-3c and STS-13c. Also supported are ATM user network interfaces (UNIs), network-to-node interfaces (NNIs), DS3, E-3, multimedia bridges, video servers, and data servers. The switch can be configured with a capacity of 2.4 to 20 Gbps, with multiple priorities for cell loss and delay. The shared memory design has been created to ensure that peak rates of bursty and circuit-oriented traffic can be carried without degradation in quality of service. The switch can be integrated with existing switched multimegabit data service (SMDS) and frame-relay networks. There's no single point of failure within this switch.

The service management module

The service management module provides integrated network management for operations, administration, maintenance, and provisioning. Based on UNIX and the BaseWorX applications platform, the module has a customizable graphical user interface with advanced open systems, object-oriented software. It supports an SNMP agent. While this module supports ATM Forum policing parameters, it also does much more with a *virtual scheduling algorithm* which enables service providers to allocate and manage service-node bandwidth efficiently. This module can be located at a service node or at a remote site.

The access module

This module is designed to provide efficient access, multiplexing, and concentration in the network, either at the service node's location or at a remote site. It can be configured to support access to most major applications, including data network services (SMDS and frame relay) and entertainment video. It supports ATM and STM in both SONET and SDH environments, as well as video multiplexing functions that will terminate ATM streams over SONET OC-3 and perform MPEG-2 transport stream demultiplexing for efficient distribution of entertainment video services.

The BNS-2000 data server

The BNS 2000 cell relay switch is the broadband data server for the GlobeView 2000 system. When used in conjunction with the access module, it supports frame relay and SMDS at the same time as ATM. BNS 2000 networks can be interconnected with one or more T-3/E-3 BNS 2000 ATM modules. SMDS is supported by using the BNS 2000 as a connectionless server.

The switched services module is designed to supply real-time call processing, signaling, and control for basic to full B-ISDN. This module offers rapid deployment of switched services, access to a full range of provisioned services, and support for integration with other intelligent network elements and existing narrowband networks. This module provides the call processing and signaling capabilities for ATM switched virtual circuits (SVCs) and seamless internetworking with AT&T's 5ESS switch. Users of a broadband network can make voice, data, video, or multimedia calls anywhere on the network and use virtually any bandwidth.

The LCS family

The LCS200 family of routers and bridges provides internetwork connectivity with LANs. These devices are essential for LAN emulation.

Configurations

AT&T offers its customers different combinations of the broadband network components just described. The idea is to match particular configurations with particular customer needs and also to create a migration path for customers as their needs change. Here are some configuration options:

GCNS: ATM high-speed data system and services. The GCNS configuration includes the service node, service management module, and access module. There's also an auxiliary maintenance module for each additional local service node as well as optional data and video servers. Designed as an ATM backbone vehicle with permanent virtual circuits (PVCs), the GCNS also supports SMDS, frame relay, and video. The BNS 2000 switch functions as the data server and provides the SMDS and frame-relay connections to the data, images, video, and voice traffic carried through the GlobeView 2000 broadband system. Figure 10.11 illustrates the GCNS configuration.

AXC: ATM bandwidth management system and services. This configuration of the GlobeView 2000 system includes the service node, service management module, and access module. AXC, designed for service providers who need bandwidth delivered in various increments on their circuits, functions as a virtual path/virtual circuit cross-connect configuration and can be used where standard SDH interfaces are required.

BSS: ATM system for switched services. The BSS configuration includes the service node, service management module, access module, and switched services module.

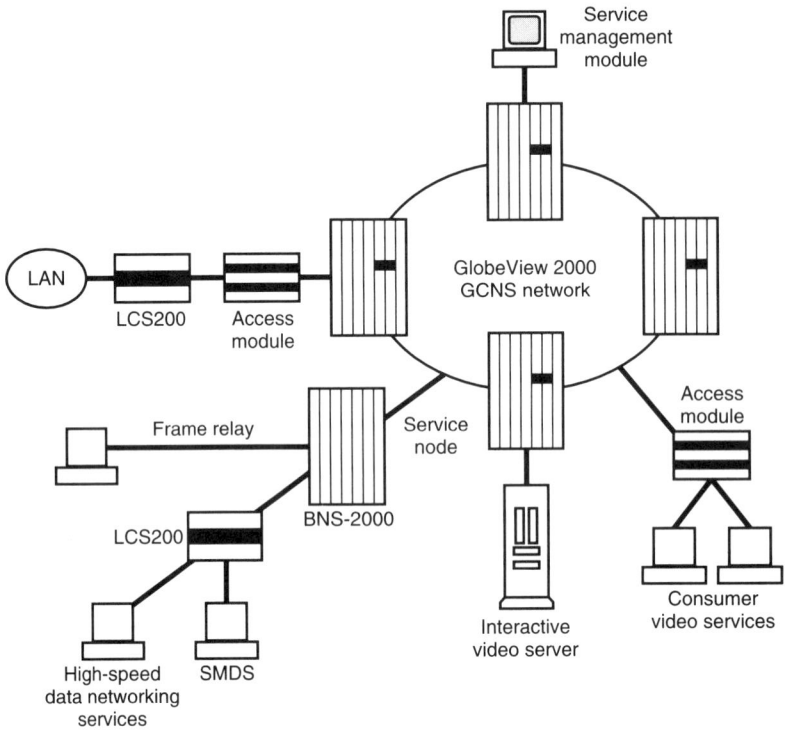

Figure 10.11 The GlobeView 2000 GCNS configuration.

This configuration is designed to preserve and extend the value of current network investments. You can upgrade the GCNS and AXC to the BSS by adding the switched services module. The BSS allows transportation of synchronous circuit traffic in an ATM format, and is capable of internetworking smoothly with narrowband networks using SS7 signaling. The BSS supports the ATM Forum's UNI and UNI/NNI signaling standards. AT&T positions this configuration as one that will provide such services as video on demand, teletraining, medial imaging, virtual reality, and multimedia telephony.

Future support for the GlobeView 2000

AT&T has indicated that it will provide the following support for the GlobeView 2000:

LAN interconnects. High-speed paths to link one LAN to other LANs

WAN distributing computing. Distributed computing power to data and video services throughout a wide area network

Interactive video server. An adjunct to the GlobeView 2000 that enables service providers to offer viewers programs and interactive capabilities

Summary

Stratacom has been in the forefront of vendors developing ATM switches that can accommodate voice as well as data and video traffic. It offers backbone (BPX) as well as network edge switches (IGX and IPX). The company has been particularly successful in attracting customers who currently use T-1 lines. Its major strategic partnerships are with Cisco Systems and Bay Networks. Newbridge Networks offers both local area network and wide area network switches. Its VIVID line of switches focuses on the local area network. The company believes that it has a competitive advantage in its routing capabilities by linking LAN and ATM switches, creating a wide area network. Newbridge Networks has a strategic relationship with Standard Microsystems (SMC) that enables it to take advantage of SMC's ability to create customized silicon for switches.

Cascade Communications focuses on ATM switches, particularly those it characterizes as second-tier carrier switches. These switches link switches at customer premises with carriers' backbone switches. It touts its Cascade 500 switch as one of the most advanced ATM switches because some functions previously performed by software are now integrated into the hardware for greater speed and efficiency.

General Datacomm Corporation (GDC) has focused on ATM technology that can be used by its large customer base who still have time-division multiplexers. It has focused on providing data-only ATM switches because it doesn't believe there's a demand for ATM switches that also handle voice. It has a strategic relationship with Digital Equipment Corporation.

Northern Telecom (Nortel) offers a number of ATM switches, including the Magellan Concorde, Vector, and Passport. Its switches range from backbone core switches for carriers to edge switches that can be used on private as well as public ATM networks. The company has a strategic partnership with FORE Systems and is incorporating FORE technology into its switches so there will be seamless LAN-to-WAN links between the two companies' switches.

AT&T offers its GlobeView 2000 system as a complete ATM network. The system includes various modules that enable customers to incorporate data servers, routers, network management, and other functions to customize their networks.

11

Madge Networks and
First Virtual: Desktop ATM 25

This chapter looks at a relatively new type of ATM application: desktop ATM. Sharp drops in prices as well as the development of some data-intensive applications such as multimedia have created a market for this product. I'll describe the phenomenon of 25-Mbps ATM as well as some of the innovative new multimedia applications. Because you've already examined IBM's products, I'll focus on two other vendors who offer this type of application: Madge Networks and First Virtual Corporation.

The Growth of 25-Mbps ATM

After initially approving a specification for 155-Mbps ATM over category-5 cabling, the ATM Forum modified this specification so it could also run over copper wire. Still, the price point for 155-Mbps ATM at the desktop is still prohibitive for most applications, which leaves ATM at 25 Mbps. After initial rejection, IBM and a number of other vendors managed to convince the ATM Forum that it should approve a set of specifications for ATM 25, a 25-Mbps version of ATM designed specifically for desktop applications. Adherents argue that it has a lower cost per port than switched Ethernet or Token Ring, as well as 100-Mbps Ethernet.

The road for 25-Mbps ATM to the desktop hasn't been easy. A number of vendors led by Sun Microsystems have argued that next-generation multimedia applications to the desktop will require 155 Mbps. Not so says the ATM 25 Alliance. This group, including First Virtual, Madge Networks, IBM, Whitetree Network Technologies, Optical Data Systems, and Xircom, believe that the public will accept the 25-Mbps version for a number of reasons:

- The ability to run over existing voice-grade, category-3, untwisted-pair wire means that around 60 percent of existing LAN wiring can support this technology.

- There should be an easy migration for companies that want to upgrade later to 155 Mbps or 622 Mbps when they need these speeds.

- The 25-Mbps speed should be adequate for current multiple video channels (MPEG 2), CD-quality audio, multimedia, and video conferencing applications.

- The low cost of ATM 25 will encourage developers to produce multimedia applications for ATM 25 networks.

- 155-Mbps desktop ATM would require substantially larger buffers than ATM 25, and these buffers will always keep the price much higher than that of ATM 25.

- ATM 25 provides five times the throughput of switched Ethernet and will attract customers when they realize this fact.

Madge Networks: Bringing ATM to the Desktop and LAN

Madge Networks has set two key goals for itself as a company: to deliver the broadest-range, best-of-class switches for ATM in LANs, and to be the number-one LAN switching company for large organizations. As you'll see shortly, the goal for the best-of-class switch extends to the desktop. In order to achieve its switching goals, the company first had to fill some gaps in its product line.

Madge Networks acquired LANNET Data Communications to take care of some serious gaps in its switch and hub lines, especially in the areas of Ethernet topology, chassis-based hubs, and hub management. Madge has been very strong in the area of Token Ring products. In fact, because of its emphasis on quality and technology leadership, it's usually considered as the one acceptable alternative by companies that are "true blue" IBM customers. LANNET, on the other hand, has developed a reputation for having some of the fastest Ethernet switching hubs in the industry. The result of the acquisition has been a concerted effort by Madge to link together the two companies' technologies and develop a plan for customers to take advantage of legacy LAN switching that includes both Ethernet and Token Ring, and then migrate these customers and their installed base of equipment smoothly to an ATM network. It has described the technology mix between the two companies as shown in Table 11.1.

Madge's vision of a switched network environment

To understand the role of desktop ATM 25 in a Madge environment, let's take a moment or two to look at the company's plan for a switched network environment. The company's architectural philosophy is as follows:

- Maximize switching and minimize routing
- Filter broadcast packets in access switches and in LAN emulation services
- Maximize the size of emulated LANs

TABLE 11.1 The Technology Mix for Madge and LANNET

Madge core technology	LANNET core technology
Desktop expertise	Chassis-based enterprise platform
Token Ring switching, including silicon	Ethernet switching, including silicon
Token-Ring-to-ATM technology	Ethernet-to-ATM technology
LAN emulation software	Enterprise network management
ATM switching silicon and software	

- Deliver ATM-like functionality on shared LANs
- Deliver end-to-end management of switched networks

Madge sees the network as eventually being a mixture of workgroup, access, and backbone switches. In the interim, though, legacy LAN switches will play a crucial role through LAN emulation over ATM. The company sees an evolution toward ATM switching that begins with Ethernet, fast Ethernet, and Token Ring workgroup legacy LAN switches. On the Ethernet (LANNET) side, the company offers chassis-based hub switching for Ethernet, fast Ethernet, and FDDI switching. On the Token Ring side, you have Madge's cut-through Token Ring switch, called a Ringswitch. The company believes that customers want a clear migration path to ATM, though many feel far more comfortable purchasing legacy LAN switches today. The result is that Madge will enable customers to migrate to high bandwidth in the future by adding ATM or FDDI modules to these switches.

Madge believes there's already a market for 25-Mbps ATM desktop switches for specialized multimedia/video applications that can take advantage of ATM's features. Its Collage 250 workgroup ATM switch is designed to bring ATM 25 to the desktop. It has a stackable architecture and can provide up to 144 desktop ATM 25 ports. In addition to meeting all current ATM Forum standards, the switch can "self-learn" and be configured for most applications. Madge's Trueview network management software enables a network manager to view the status of all Madge devices on the network, as well as configure and perform remote actions from a single console.

Madge's Collage 280 is a workgroup ATM switch that permits *adaptable switching* between Ethernet and ATM; for network managers, this might be the ultimate in flexible desktop switching. By changing the adapter card in an attached PC, the switching technology of the Collage 280 allows each of the 12 desktop ports to automatically adapt to stacked Ethernet or ATM 25. Adaptable switching is a feature that's bound to be near the top of the list for many customers who want to be able to decide when they're ready to switch to desktop ATM. As a member of the Collage family, this switch is scalable to support up to 144 desktop ATM 25 or Ethernet ports.

While desktop ATM switches are the focus of this chapter, note that Madge has plans for a full line of Collage ATM access switches for Ethernet-to-ATM (the Collage 530) or Token-Ring-to-ATM (the Collage 540) access. These switches will take advantage of a high-speed ATM backbone that allows connectivity between the Ethernet and Token Ring ports and a 155-Mbps ATM port. Network management will be provided by LANNET's MultiMan suite of network management applications. Finally, the company will offer its own backbone ATM switch, the Collage 740. It will support up to 16 ports of 155-Mbps ATM as well as a variety of other interfaces.

Voice as a future desktop ATM application

Madge's vision of the future sees desktop ATM's first applications as primarily multimedia and video, but it also believes that voice can play a role. The first desktop voice applications are likely to be specialized applications where voice and data integration is important, such as automated call distribution in customer service centers and interactive voice response systems. A little further in the future, however, Madge sees a PC with an ATM 25 connection as the voice terminal for general-purpose desktops. Voice, e-mail, fax, and video would be integrated as a single PC-based application. Telephone calls could then incorporate video conferencing rather easily. The telephone closet would look quite different than it does today. Voice-mail servers, interactive voice and response servers, remote access servers, and automated call distribution servers would transmit data to an ATM 25 switch. This switch would then communicate the information to an ATM 155 switch that would send it to file and print servers, application servers, and distributed ATM 25 switches. These switches would bring integrated voice, data, and video to ATM end stations on people's desks.

Should Madge Networks be your ATM vendor?

Madge Networks has built its reputation primarily on Token Ring topology, but it's known for its state-of-the-art technology as well as its ability to design and manufacture its own silicon products. Its purchase of LANNET appears to be a very nice match, and the company's evolutionary plans for ATM seem to do a good job of melding the two product lines. The company does understand switching and has been very active in the development of ATM LAN emulation software. One area it has not been a leader in, however, is the virtual LAN arena. But this should change when its Collage family of switches begin to appear.

Madge Networks appears to be the logical choice for a company with a Token Ring environment who wants to migrate eventually to ATM. For companies with both Token Ring and Ethernet (over half of the largest Fortune 1000 companies in this country), Madge's acquisition of LANNET makes it a leading contender for the vendor of choice in such a mixed environment.

For companies that are Ethernet-centric and have no Token Ring equipment, Madge is probably an unknown quantity. Coming from the Token Ring world, the company places its emphasis on advanced technology and solid support rather than on offering the lowest prices in the industry. As a result, Madge won't win every competitive bid, but the company is likely to do a good job as the ATM vendor of choice.

Multimedia as the Killer Desktop ATM 25 Application

Multimedia has clearly been identified as the type of application that will drive the sales of desktop ATM. There has been considerable debate about exactly what type of bandwidth is needed for multimedia applications. Table 11.2 shows video throughput requirements for a number of different applications.

TABLE 11.2 Video Throughput Requirements

Application	Service	Bit rate
Videophone	H.261/H.320	64-Kbps to 2-Mbps CBR
CD-ROM	MPEG-1	1.5-Mbps CBR
	MPEG-2 "VCR quality"	4-Mbps CBR
CATV	MPEG-2 "TV quality"	10-Mbps to 15-Mbps VBR

Early desktop video applications over ATM will likely take a number of different forms. One practical use of real-time video would be collaborative applications that are shared by video conference. A second real-time video application is likely to be customer consultation with remote experts. Stored video could be accessed via desktop PCs for training, to view news broadcasts and company communications, and to view point-of-sale promotions.

Why Desktop Multimedia Requires ATM

A desktop multimedia application such as PC video conferencing cannot tolerate much of a delay between the time a user sends a message and the time a reply is received. First Virtual Corporation characterizes a delay greater than 200 milliseconds as annoying and a delay of more than 400 milliseconds as intolerable. A second problem for an application such as video conferencing is that, when it's transmitted over shared media such as Ethernet, there's a random variability in the delivery of audio and video packets because of the bursty traffic associated with this type of network topology. In some cases, audio and video streams might not be synchronized.

The answer appears to be desktop ATM. What is very appealing about ATM technology for PC video conferencing is that it offers bandwidth on demand as well as guaranteed quality of service. Bandwidth can be reserved on the basis of priority.

First Virtual's Vision for Multimedia Over ATM

First Virtual has developed a 25-Mbps ATM switch, multimedia server software, and a multimedia operating system to go along with PC-based ATM network adapter cards. Ralph Ungermann, the CEO of this new company, sees the goal of his company to be the development of technologies that allow network managers and software architects to build affordable multimedia applications for departmental users. Initially, Ungermann thinks it's important that the ATM structure support Ethernet and integrated services digital network (ISDN). Ungermann has stated that his company is "selling an application, not an infrastructure," which makes the fact that the technology runs on ATM almost incidental. He believes that his customers are "putting ATM in for the application." To understand how First Virtual provides ATM support for multimedia applications, let's examine each of the key elements of a First Virtual system.

The media storage server provides store-and-forward streaming services over ATM LANs for up to 50 users simultaneously. It ties into ATM workgroups via a 100-

Mbps direct interface to ATM networks. Multiple servers can access this server, and it supports real-time storage and retrieval of multimedia formats, including MPEG, M-JPEG, and Indeo. It also provides transparent access to audio and video files from Windows and Video for Windows. It comes with First Virtual's media operating system (MOS) and 6 GB of RAID storage. The company has synchronized disk access, physical disk organization, controller and bus speed, and network access in order to provide the maximum number of uninterrupted multimedia streams. Figure 11.1 shows a typical First Virtual multimedia ATM system in action.

The media switch is designed to distribute multimedia applications among workgroups. Because it offers point-to-multipoint service, it's optimized for multimedia applications, including e-mail and video conferencing. It supports the ATM Forum's quality-of-service specifications, which are software selectable so that different priority queues can be allocated for each outbound port. The switch supports ATM 25 and has eight full duplex 25-Mbps ATM ports for the desktop and one LAN emulation port. It also has two 100-Mbps TAXIs for higher-speed connections. It's possible to stack as many as five of these switches for an aggregate capacity of 5 Gbps of throughput.

The First Virtual switch supports both permanent virtual circuits and switched virtual circuits. It also has a fast signaling protocol that has been optimized for multimedia so it can dynamically establish SVCs among attached network devices. Because the switch is designed to coexist in a legacy LAN environment, it has an Ethernet transparency module (ETM) that enables NetWare, NetBIOS, and other TCP/IP applications to communicate seamlessly over an ATM network.

The media operating software (MOS) runs under the application and above the network operating system. It provides the basic interactive network service for video and telephony services. In effect, it acts as ATM middleware to link together all parts of First Virtual's multimedia system. As Figure 11.2 illustrates, the MOS can provide the video and telephony services to workstation PCs in a LAN environment.

One significant early application of MOS is MOS for Notes. By adding support for both Lotus Notes and Lotus Video for Notes, First Virtual believes it has a winner because it makes it possible to transmit Notes on ATM right out of the box. To support

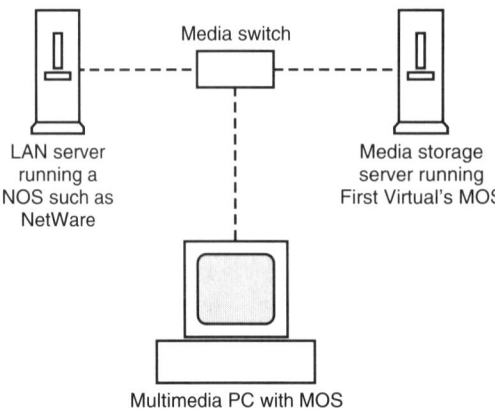

Figure 11.1 A typical First Virtual multimedia ATM system in action.

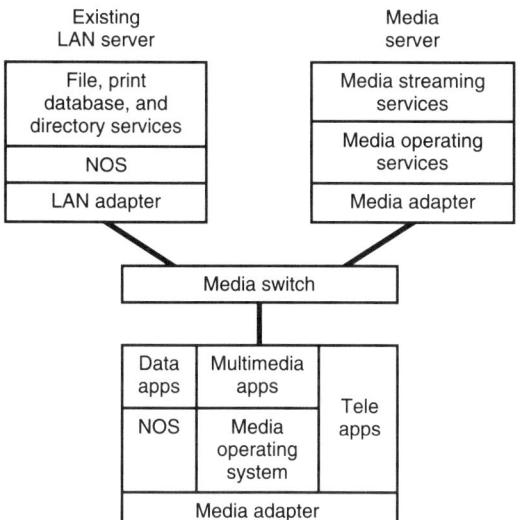

Figure 11.2 Existing LAN servers and First Virtual's multimedia ATM system.

Notes, First Virtual had to add separate MOS agents for the Notes server, the media server, and client software for the end user's PC. This product makes it possible to create scalable groupware and multimedia applications that integrate information taken from video, data, and image databases. Video streams can be replicated to other Notes servers over a router-based network. Among the first Notes applications being developed that use multimedia and First Virtual's system are workgroup training programs. A major advantage of running Notes over First Virtual's ATM links is that you can take advantage of ATM's quality of service and classes of service attributes.

What will probably make MOS attractive to companies planning multimedia applications are the functions it allows, among which are the following:

- The ability to play back, record, edit, and direct multimedia objects and files over an ATM network

- The ability to record video in real time and save the files to First Virtual's media storage server

- The ability to transfer compressed digital video files to multiple users

- The ability for managers to manage sessions, files, and objects from a standard Windows File Manager application

- The ability to embed a multimedia file into another application using object linking and embedding

Media adapters

These adapter cards support Microsoft's NDIS and Novell's ODI drivers, as well as several network operating systems. Models support 25 Mbps over category-3 or category-5 unshielded twisted-pair wire. There's a model that supports 100 Mbps over multimode fiber. There's also an optional daughtercard that provides MVIP (multivendor interface protocol) support, which enables video to pass directly from adapter to codec in order to bypass potential bottlenecks in the PC bus.

ATM and ISDN for video

First Virtual has combined standard H.320 video conferencing with ATM and integrated services digital network (ISDN). The company interfaces H.320 video conferencing systems from AT&T or PictureTel directly to an ATM network via a codec and an ATM adapter for PCs. The codec supports AT&T's Vistium 1200/1300 and PictureTel's Live PCs 100 video conferencing systems.

First Virtual's MOS software links the video conferencing application in the PC to the ATM network. It's part of a H.320-compatible media gateway server. The gateway can support multiple basic rate interface (BRI) or primary rate interface (PRI) lines. This support for multiple lines allows First Virtual's system to provide transmission rates up to 384 Kbps per session, so a video quality of 30 frames per second is possible.

The advantage of this gateway is that a workgroup can share data, video, and graphics with remote participants without having to install ISDN connections on each desktop. Each switching adapter for a PC's ISA or EISA bus includes three 25-Mbps ports to connect to the codec, the ATM switch, and a multivendor integration protocol voice bus.

Should First Virtual be your ATM vendor?

First Virtual has had several firsts for such a young company. It has figured out a way to run ATM over category-3 wire, how to run multimedia applications over ATM, and a way to link to legacy LANs. Its products currently fall into a gray area where there are no real standards. The company has a ready answer for potential customers who worry about being locked into proprietary systems. The fact that First Virtual's ATM implementation is based on software rather than hardware means that it can accommodate evolving standards reasonably easily.

Give First Virtual a lot of credit for some very innovative technology. Rather than tie up capital with its own manufacturing, the company has chosen to develop several key strategic relationships with other vendors to leverage their manufacturing capacity. In fact, the only strike against this company from the perspective of some potential customers is that it's still a very new player and customers are concerned about its long-range chances for success. Supporters point to Ralph Ungermann, the company's CEO, as the steady hand who has already built one giant networking company.

Summary

There has been a good deal of debate over the merits of 25-Mbps versus 155-Mbps ATM at the desktop. The argument for ATM 25 focuses on its attractive price point, its adequacy for many video and multimedia applications, and its scalability to higher speeds if required.

Madge Networks is likely to be a major player in the ATM 25 desktop marketplace. It currently offers legacy LAN switches for both Ethernet and Token Ring. These switches can be migrated to ATM or FDDI through the addition of an ATM or FDDI module. Its Collage family of switches includes several desktop ATM 25 products. Some feature ATM adaption, the ability to adapt to ATM as well as to Ethernet or Token Ring simply by installing the appropriate adapter card in the PC connected to the switch. Madge foresees a time when integrated voice, data, and video can be found at the desktop via ATM 25.

Multimedia that includes video is likely to emerge as the "killer app" for ATM 25. First Virtual Corporation is in the forefront in this area and offers its own middleware software that stands between the application and the network operating system. Its media storage server, multimedia operating system, and ATM 25 switching functionality are sold as one system. In fact, CEO Ralph Ungermann believes that the key to success with desktop ATM 25 is to sell a multimedia application and include the system as just something required to run it. In other words, the emphasis should be placed on the application and not on the ATM technology.

Evaluating ATM

12

Evaluating the Enterprise
Network Environment for ATM

Where do you start when you decide that you probably need ATM? The purpose of this chapter is to provide a basis for examining your current enterprise network environment much the same way a high-paid consultant might. There are several checklists of questions to ask yourself as well as prospective vendors, along with explanations of the types of responses you should grade favorably. Let's face it; all vendors' literature extols their features and cleverly ignores services they don't provide. Which features are really important and which are irrelevant for your particular needs? If this chapter can help you answer some of these questions, then it will certainly more than justify the price of this book. So roll up your sleeves and get ready to take a closer look at your network environment and the current state of enterprise switching products.

Examining Your Current Enterprise Network

Before considering ATM equipment, it's absolutely crucial that you take a good long look at your current equipment as well as your future business needs. Here are some topics to be sure to think about.

The changing business environment

Your business requirements (present and future) should drive your networking plans. If new applications will include multimedia and be data traffic-intensive, then clearly something must be done to ensure adequate bandwidth and avoid traffic congestion. If the company plans to expand its enterprise network to include units that have different computing platforms, then the question becomes whether or not these platforms can be made interoperable with your current networking equipment. If future plans call for more branch offices, then LAN-to-WAN communications and wide area ATM specifications become far more important considerations.

Dead-end topologies

Does your company have any "dead-end" topologies, such as Arcnet or LocalTalk? While there's a 20-Mbps version of the 2.5-Mbps Arcnet, for all practical uses this topology is dead. Similarly, the 230-Kbps LocalTalk isn't going to be adequate for any kind of sophisticated network application in the future. So one key question you must ask is "How can I modify these topologies?" You can try installing cheap Ethernet cards and then use Ethernet switches to effectively segment according to workgroup needs.

Wide area network links

Do you have branch offices that will need to be part of any enterprise network? If so, the idea of seamless ATM local-area-to-wide-area network transmissions will sound very appealing. Wide area network links mean that you have to determine what WAN interfaces you'll need. If you currently use frame relay for your wide area network, then the new specifications for frame relay to ATM might be very appealing.

One area of concern should be the lack of interoperability among wide area network switches. If your company wants to incorporate LAN-to-WAN links using ATM, then you'll probably have to focus on companies that have made efforts to test their switches for interoperability or who have developed strategic partnerships. If you're looking at Digital switches on a LAN, you might also want to look at GDC, its strategic partner. Similarly, a company interested in FORE System switches should investigate what Nortel has to offer. The other solution to this problem is to look at a few vendors, such as IBM, that profess to offer the complete LAN-to-WAN ATM solution.

What Size ATM Switches Are Desirable?

Some network managers have a problem that's of their own creation. They try to save a little money up front and end up selecting the wrong ATM switch for the job. A workgroup switch can't be expected to perform like a campus switch, and a campus switch just isn't adequate as an enterprise switch. This fact is particularly true because some ATM switches will drop cells when there's congestion and no buffer space left.

Workgroup ATM switches are ideal if the major purpose for adding ATM is to increase bandwidth to a limited number of desktops at a specific location. Today, desktop ATM means 25-Mbps PCs running real-time multimedia applications and 155-Mbps high-performance workstations running scientific and engineering applications. Workgroup ATM switches generally handle fewer than 10,000 virtual circuits and don't come with wide area network interfaces.

Campus ATM switches are ideal if the primary role is to support links among legacy LANs. These switches convert Ethernet and Token Ring packets to ATM cells and can handle up to a couple of dozen legacy LANs. They generally come with limited numbers of wide area network interfaces, and can handle in the range of 10,000 to 20,000 virtual circuits.

Enterprise ATM switches play the role of interconnecting all campus ATM switches as well as providing the wide area network backbone for a private network. They support over 20,000 virtual circuits and several different wide area network interfaces.

Selecting the Right Virtual LAN

As discussed in chapter 3 in general and chapters 5 through 11 in terms of specific vendors' products, virtual LANs are currently in an embryonic stage of development. There are no real standards as yet, although Cisco has actively pursued some. There's little agreement about the level of virtual LAN segmentation, the way that such networks can be managed, and the interoperability of such networks. A network manager considering legacy LAN and ATM switches needs to consider whether the virtual LANs created from such switches meet the needs of the company. Companies with networks that contain multiple protocols might have a real need for level-3 routing to group users by protocol type. Network management is another major differentiation among virtual LAN products. One of the key questions that needs to be asked is whether or not the virtual LAN can be managed in conjunction with all components of the enterprise network.

As part of this management, vendors who offer or promise policy-based management of their virtual LAN products should be given serious attention. Policy-based management lets network managers assign priorities to different types of application traffic. This means that higher priorities can go to core business applications. If an outage narrows the amount of network bandwidth available, then policy-based management ensures that these crucial applications receive the necessary bandwidth. Cabletron is committed to policy-based management, as is Digital Equipment Corporation.

The Multifunction Intelligent Hub

Many vendors tout their intelligent hubs as the focal point for their enterprise networks. Far more is demanded of today's intelligent hubs than earlier models. If your company is looking at adding intelligent hubs, there are several features you can use to differentiate the hundreds of products available.

One key question that any company considering an intelligent hub must ask is whether it prefers stackable or chassis models. Stackables offer many advantages. They tend to be far less expensive and very flexible. Companies looking for an enterprise-wide network management scheme might want to consider chassis-based hubs because they usually offer more sophisticated network management functions. They also tend to support a wider range of network topologies and offer very high-speed backplanes. The best of both worlds might well be Digital Equipment Corporation's stackables, which can be combined to form a chassis architecture when the features associated with a chassis are needed. You might also want to ask some additional key questions:

What interface modules are available? Intelligent hubs are easiest to differentiate in their support for high-bandwidth topology. Not all vendors offer FDDI, for example, or 100-Mbps Ethernet. Some offer limited fiber support, so you couldn't have, for example, fiber-based Ethernet or fiber-based Token Ring LANs. Some products no longer support coaxial cable, while others support only shielded and not unshielded versions of twisted-pair wire interfaces for Token Ring LANs. Not all products support AppleTalk LANs, but your company might already have it. In other words, take a close look at your current network environment along with future business plans and determine what topologies and interface types need to be supported.

What is the total bandwidth available? Depending on your current and projected needs, some intelligent hubs might not have the bandwidth you require. Look at the difference, for example, between Alantec's PowerHub 7000 with its 3.2-Gbps bandwidth and UB Network's GeoLAN 500 and its 40.4-Gbps bandwidth. Which is required and which is overkill? Only you can decide based on current and future plans your company has for its network use.

Do you need mainframe support within your hub? Some companies such as Cabletron have focused on providing IBM SNA support for its intelligent hubs, while other vendors ignore this need.

What types of legacy LAN switching are supported? This can be a significant differentiation. Some smaller vendors support only Ethernet switching, while others are now adding Token Ring and even FDDI switching. For companies that have a large number of workgroup-centric LANs, the intelligent addition of legacy LAN switches can offload traffic from overtaxed backbones and delay the need for changes in the backbone's infrastructure.

What is the intelligent hub family's architecture? What future growth and migration paths are available with this architecture? These questions can reveal significant gaps in vendors' product lines. 3Com, for example, has been slow to embrace ATM, has argued repeatedly for the continuing importance of routers, and has chosen to go the stackable path rather than sell a chassis model. Digital Equipment, on the other hand, has embraced both FDDI and ATM switching along with legacy LAN switches and offers stackables that can be grouped into a chassis whenever desired. Some hubs have data buses that support a fixed number of shared media LAN segments. Only certain slots can be used for switching modules. Other hubs offer undifferentiated slots that can be used for switches or legacy LAN modules.

Are terminal and communications servers supported within the hub? Companies with mid-range systems and branch offices or mobile work forces need to find the answer to this question.

Is routing supported within the hub? The answer to this question will reveal vendors' philosophical as well as their product differences. Some companies, such as Fibronics and 3Com, argue for a full-featured router in front of the intelligent hub.

What role will ATM play in conjunction with the vendors' architectural plans? Some vendors plan to include switching modules within their hubs, while others plan to place stand-alone switches in front of their hubs. Where the translation takes place between cells and frames is important. Switching modules are likely to be less expensive than stand-alone switches and interoperability isn't an issue because the vendor provides the integration.

What ATM interfaces will be available within the hub? Some vendors, such as IBM, envision the intelligent hub as the focal point for desktop ATM links, as well as links to backbone ATM switches and workgroup ATM switches. Other vendors are planning to support only 155-Mbps interfaces.

Can existing hubs and/or their modules be used in the vendors' new network vision? Some companies, like Cabletron and UB Networks, permit their customers to use their present equipment as part of their future growth. Bay Networks, on the other hand, hasn't been able to roll back features found in its newest products to its installed base of System 3000 users because of the limitations of that hub's backplane.

Legacy LAN Switches

Configuring them for best results

- Use an analyzer before installing the switch, and allocate switch ports based on the analysis. Unfortunately, only a few switches support RMON MIB, so that might be a legitimate criteria for selecting a switch.

- For multimedia, look to cut-through switches to keep delays low and constant.

- A good rule of thumb is to keep traffic on each switch segment to about 50 percent.

- With switched Ethernet, make sure that servers and printers are on the same switch as the users accessing them.

Questions to ask

If it's a CPU-based device, how does it function under heavy loads? Some switches' forwarding rates drop alarmingly when loads increase.

Once again, if the switch is CPU-based, what percentage of frames are dropped under heavy loads? Many companies are beginning to move from CPU-based to ASIC-based switches to alleviate this problem.

Is the processing scheme store and forward or cut-through or a combination of each? In chapter 1 I discussed the advantages and disadvantages of each of these approaches.

Is the backpressure feature present? This flow-control feature helps control frame loss. Each time a buffer on a switch port starts to fill up, the switch fools the devices attached to that port by sending a signal indicating a collision. This causes the devices to pause and back off before trying to transmit data again.

Are high-speed ports included? Many switch vendors are beginning to add 100-Mbps Ethernet and FDDI ports to their switches.

Does the Ethernet workgroup switch operate at the network layer? Most of these switches operate at the MAC layer and aren't capable of establishing firewalls.

A legacy LAN switch checklist

- What are the maximum number of switched ports offered? How is expansion accomplished?

- What is the price per port?

- What kinds of WAN connections are available?

- What is the maximum throughput of the backplane (in Mbps)?
- What protocols are routed?
- What filtering criteria are supported?
- What forwarding method is employed?
- What are the maximum number of MAC addresses on each port?
- Is SNMP management supported?
- Is SNMP management supported with the switch?
- Is the RMON MIB supported?

Major Issues Concerning ATM Switches

Purchasing ATM switches is a very complex task. Following are some key issues when considering specific vendors' products.

Switched virtual circuits (SVCs)

The standards for switched virtual circuits are becoming more stable, but vendors still vary widely in what they offer. The two generally incompatible SVC signaling specifications are the ATM Forum's user-to-network interface (UNI), versions 3.0 and 3.1. Vendors have been shipping products with proprietary SVC signaling, but should be moving toward 3.1.

LAN emulation

The two standards currently in place for LAN emulation are the internet's request for comments (RFC) 1577, known as *classical IP over ATM*, and the ATM Forum's LAN emulation 1.0 (LANE). Unfortunately, there's still a lot of proprietary LAN emulation used by products in the marketplace.

Network management

There are still a lot of gaps in network management offerings for ATM. Most equipment vendors do offer basic network management features, including automatic rerouting around network trunk failures, fault management, and point-and-click configuration. All vendors say they offer easy-to-use graphical interfaces.

Simple network management protocol (SNMP) support should be available from all the major equipment vendors; some are even planning to offer support for common management information protocol (CMIP). Network management features become even more crucial in core switches and central office switches because of the vast number of circuits that must be managed. Network modeling is available under such network management systems as Cascade's CascadeView and StrataCom's Strata-Sphere. A StrataCom HealthChecker module automates optimization and modeling functions, including the provisioning of permanent and switched virtual circuits.

One of the major thrusts of companies that sell switches to carriers is including intelligent networking capabilities with network management. This means the development of features built around the processing, billing, and routing of calls that are based on factors such as the time of day, caller ID, receiver ID, or other information that can be captured in a database and used to process a call.

Know what *nonblocking* means

Look at the throughput on a switch and then look at the number of ports to determine if it's really nonblocking. If a switch has 2.4 Gbps of throughput and 16 ports, for example, each port can run at a full 155 Mbps. Nonblocking means that, when a cell appears at an input port, it will be switched immediately to the proper output port. Unfortunately, it doesn't mean that cells won't be lost. If an outgoing circuit is congested, cells could still be lost even though they aren't blocked.

It's important to realize that, while all matrix switches by their nature are nonblocking, other switches can still be attractive candidates because the data traffic your network generates might not need the nonblocking feature. If a switch can handle only 1.2 Gbps of throughput, for example, and you know that your data traffic will never exceed 1 Gbps, then for all practical purposes the switch is nonblocking for your environment.

ATM Network Interface Card Checklist

What bus architectures are supported? SPARC and Sbus are supported for high-performance workstations, but you're beginning to see added support for PCI as well as EISA. There's also scattered support for Micro Channel and ISA architectures.

Are both SVCs and PVCs supported? There are still vendors with NICs that support only PVCs.

What ATM interfaces are supported? There's a wide range of support here, ranging from ATM 25 UTP to OC-3 (155 Mbps). In some cases the TAXI interface might exhibit the shortest delays of any of these interfaces because it's a native cell-bearing interface that provides direct access to raw ATM cells. Over the OC-3 interface, on the other hand, ATM cells must be segmented into and reassembled from SONET frames.

What network operating system platforms are supported? There's beginning to be more and more Windows NT support. There's also scattered support for Solaris and SunOS, and even the AppleTalk protocol.

Is there an ATM application programming interface? As shown in chapter 2, this is the long-term goal for optimum use of ATM.

What flow control is supported? Is it the ATM standard? Vendors use different methods of flow control to handle congestion. If your company is interested in installing a multiple-vendor ATM network, then this area is crucial.

Is there an SNMP agent on the adapter? You want to be able to manage the adapters as well as the rest of the network.

What does the NIC device drivers contain? The features built into the device driver affect performance. A device driver that includes support for statistics generation, flow control, signaling, and network management will reduce ATM performance several percentage points over performance produced by a bare-bones version of that device driver. The key is to determine if a device driver contains the features you need and not additional bells and whistles that reduce performance.

What is the actual price of the adapter with the features you really need? By dividing the price of the interface by its performance (in Mbps), you can calculate the price per megabyte of performance. Unfortunately, this number can be deceptive if the vendor sells, as optional add-ons, items such as LAN emulation, the user network interface (UNI), or network management—all features that you might very well require.

Are reduced instruction set (RISC) processors on the adapters? RISC chips are often installed on adapter cards to off-load cell segmentation and reassembly (SAR) functions from the host and improve adapter-to-system memory throughput.

ATM Switches: A Checklist

What is the price per port? While most expensive isn't always best, the lowest price per port could mean the absence of a number of key features. Still, this information should be on your checklist.

What is the number of ports? Some vendors produce a single-switch model with a fixed number of ports and argue that it fits all needs, while others offer models that permit a range of ports. Flexibility is always desirable.

What's the fastest LAN interface? Most workgroup switches offer 155 Mbps. FORE has a 622-Mbps interface. Can you conceive of connecting your workgroup switches to a 622-Mbps ATM backbone? If so, look for the faster interface.

What's the maximum buffer size? As the chapters on specific vendors illustrated, there's a wide range of buffer size among ATM switches. Similarly, there's an equally wide range in the methods used to manage these buffers.

What are the number of priorities for bandwidth management? This is one area where the more you have the better, assuming you have applications that can take advantage of priorities.

Are switched virtual circuits supported? There are still a large number of vendors who promise SVCs but have not yet delivered.

What's the switch transit delay? There is a significant range among switch products. Companies that need low latency for multimedia applications should compare numbers carefully.

What's the level of redundancy? Does it include the switching fabric and the power supply? Are modules hot-swappable? Clearly these questions are more important for core ATM switches and central office switches.

Are RISC chips or ASICs used? More and more vendors are adding processing power to modules in order to maintain switch performance, even under heavy loads.

What protocols are supported? While most switches will support NetWare's IPX, some also support DECnet, AppleTalk, etc.

Glossary

application layer The OSI layer designed to handle requests made by an application program running on a network.

Arcnet A noncontention network topology with a bandwidth of 2.5 Mbps.

ARIES network The ATM research and industrial enterprise study (ARIES) is a network established by Amoco and 17 communications vendors to create an end-to-end ATM internetwork.

asynchronous transfer mode (ATM) A cell-based high-bandwidth switching technology originally designed to be the transport mode for broadband ISDN, but now viewed as the emerging high-bandwidth solution for LAN backbones, high-performance workstation workgroups, and wide area networks.

asynchronous transfer mode inverse multiplexing (AIM) An approach to linking ATM equipment and services to as little as a T-1 ATM link via an AIM interface.

ATM adaption layer (AAL) An ATM layer that provides segmentation and reassembly of data. AAL 3/4 is the ATM adaptation layer protocol developed to support a carrier's switched multimegabit data service (SMDS) connectionless service.

ATM circuit steering management information base (ACS MIB) A database that will enable analysis tools to work in an ATM environment.

ATM Forum An industry organization composed of over 700 vendors and end users that's developing specifications for the computer industry.

ATM layer An ATM layer responsible for providing routing information for cells in the form of VPI/VCI values.

automated VLANs The ultimate direction of VLANs, a type of virtual LAN that's created almost completely automatically.

automatic load-sharing A switching feature that balances loads to maximize efficiency.

automatic network routing A source-routing algorithm that moves control information through an IBM switched network by concatenation of labels found at the front of each packet.

auto-negotiation A technique used in 100Base-T networks in which network adapters and hubs differentiate the speed of transmission as well as between half- and full-duplex transmissions.

available bit rate (ABR) An ATM traffic protocol that uses excess bandwidth and network management algorithms to evaluate network congestion and eliminate cell loss.

backbone LANs   LANs that function as high-speed communication highways to connect different LAN segments.

back pressure   A technique switches use to detect that data is overloading a specific port and generate a collision detection signal to fool the port in thinking that there has been a collision so the port backs off and forces nodes and segments to wait before retransmitting data.

basic rate interface (BRI)   A single access point into ISDN that consists of two 64-Kbps bearer channels and one 16-Mbps data channel.

bridge   A network device (hardware or software) that segments a LAN based on the network addresses of the nodes.

broadband network services (BBN)   IBM's original ATM architecture plan, now a part of switched virtual networking.

broadcast server (BUS)   Under ATM LAN emulation, software that transmits broadcasts as well as multicast packets to all emulated LAN clients.

brouter   A hybrid router and bridge that bridges wherever it can and routes where it can't.

cell loss priority (CLP)   A single-bit ATM field that indicates a cell's loss of priority. A cell with this bit set is discarded when congestion conditions occur.

central office (CO)   AT&T's term for its point of presence.

channel service unit (CSU)   A digital modem that also provides some line conditioning and diagnostic functions.

CiscoFusion   Cisco Systems' architecture for switched internetworks.

client/server   A network architecture in which client software initiates requests, and then server software processes the request on a network server and transmits the results to the client.

collapsed backbone   A backbone that's "collapsed" into the backplane of an intelligent hub, which assumes the backbone's function.

connectionless network   A network in which speed is paramount, and end users have their own error-checking and flow-control software and don't rely on the standards built into the OSI model.

connection-oriented network   A network where once a connection is made, it's reliable and will continue to be so without having to keep transmitting the destination and source addresses.

constant bit-rate traffic (CBR)   An ATM protocol designed to handle a constant bit-rate traffic between two points.

contention network   A network where nodes compete for access to the shared bandwidth.

credit-based ATM traffic management   A traffic management scheme based on the advertisement of unoccupied buffers.

cut-through switching   An approach that enables a switch to forward packets without having to wait for the reception of an entire packet.

data link layer   The OSI model layer responsible for flow control and error checking.

data service unit (DSU)   A digital modem that doesn't provide any line conditioning or diagnostic functions.

dedicated Token Ring (DTR) connections   A method for increasing bandwidth to nodes generating heavy traffic using a full-duplex mode so the nodes can receive and transmit data simultaneously.

demand priority access method (DPAM)   The medium access protocol layer of 100VG AnyLAN technology.

departmental LANs   LANs that perform key departmental applications such as accounting.

distributed processing   A LAN environment where users' microcomputers perform their own processing rather than depending on the centralized processing of larger computers.

dual leaky buckets   An ATM Forum standard for traffic policing in which one "bucket" monitors average bandwidth, while the second "bucket" monitors peak bandwidth.

dynamic host configuration protocol (DHCP)   A protocol developed for Windows 95 that automatically assigns IP addresses to nodes on startup.

dynamic IISP   A dynamic version of the ATM Forum's interim interswitch signaling protocol that allows switches to automatically establish connections and pass reachability data among themselves to eliminate the need for manual specification of connection paths.

early packet discard (EPD)   A FORE Systems feature that specifies if the total number of cells in the output buffer exceeds a user-established threshold, new packets are prevented from entering the buffers.

edge router   A device that facilitates communications between ATM switches and legacy LANs by routing traffic between virtual LANs and ATM switches.

enterprise networks   Networks that encompass a company's various platforms, including mainframes, LANs, minicomputers, and high-performance workstations.

enterprise virtual intelligent switched networks (enVISN)   Digital Equipment Corporation's comprehensive architecture and plan for the evolution of scalable virtual networks.

error-free cut-through   A switching technique used by Cisco's Kalpana switches that enable them to read address headers and error check information, and forward the packets without having to wait until the entire packet has been read.

Ethernet   The LAN topology using carrier sense multiple access, with a collision detection network access scheme that transmits data at 10 Mbps.

fast-packet technology   A technology that causes a multiplexer to generate fast packets designed for a single channel only by allocating bandwidth on demand and including the destination address on the packets to speed up handling.

Fibre Channel   A very high-speed fiber-switching technology first implemented on mainframe and minicomputer peripherals that's now being positioned as a LAN backbone technology to compete with ATM.

fiber distributed data interface (FDDI)   A noncontention network with a bandwidth of 100 Mbps over fiber or twisted-pair wire (CDDI).

filter   The process in which a bridge sends a packet with a local segment's address to that segment and doesn't forward it on to another LAN segment.

forward   The process in which a bridge transmits a packet to a connecting LAN segment if the packet's destination address isn't on the local segment.

fractional T-1   A portion of a T-1 trunk's bandwidth that a customer leases.

frame relay/ATM PVC network interworking   Specifications for providing connections between frame-relay networks and ATM networks.

frame relay/ATM service internetworking   A set of specifications for converting frame-relay traffic into ATM traffic and back again.

frame-relay technology   A data-link protocol that defines variable-length data frames with only 48 bits of overhead, and provides much greater efficiency than X.25 packet switching.

frame user-to-network interface (FUNI)   A set of specifications that supports ATM data at speeds from T-1 down to 56 Kbps, and can carry a payload of up to 4,096 bytes with only 1% overhead.

full-duplex Ethernet   Ethernet with 10-Mbps channels, one for receiving data and one for sending data in a point-to-point configuration.

generic flow control (GFC)   The four-bit ATM cell-header field used across the UNI to control traffic flow and prevent overload conditions.

header error check (HEC)   An eight-bit cyclic redundancy code that's calculated over all fields in the ATM header.

IEEE 802.3   The IEEE Committee specifications for Ethernet.

integrated services digital network (ISDN)   A set of standards for carrying integrated voice, data, and video information over the public switched network.

interim local management interface (ILMI)   An ATM UNI that uses SNMP to transmit local signaling management data across the UNI.

International Telecommunications Union (ITU)   An international UN organization that produces technical, operating, and tariff recommendations for telecommunications.

internetwork operating system (IOS)   Cisco Systems' software for switched networks.

IP host-list VLANs   A virtual LAN that's segmented on the basis of nodes' IP addresses.

LAN emulation address resolution protocol   Under ATM LAN emulation, the protocol used by an end station to resolve a LAN address to an ATM address.

LAN emulation client (LEC)   Under ATM LAN emulation, all devices that communicate with devices attached to traditional LANs.

LAN emulation configuration server (LECS)   Under ATM LAN emulation, software that can respond to a query for the address of the appropriate LAN emulation server with which it can register.

LAN emulation server (LES)   ATM LAN emulation software residing on any ATM device on a LAN and responsible for resolving MAC and ATM address mappings for all emulation LAN clients.

LAN emulation user-to-network interface (LUNI)   This set of protocols fools a LAN into thinking it's communicating with an Ethernet or Token Ring LAN.

latency   The amount of time it takes for a switch to receive a packet on one port and transmit it on a second port.

local area network (LAN)   A communications network used by a single organization over a limited distance that permits users to share information and resources.

local area transport area (LATA)   The geographical area services by a local exchange company (LEC).

local exchange carrier (LEC)   The local telephone company that provides service from a customer's premises to their closest central office.

logical link control layer (LLC)   The protocol layer within a network interface card's chip that's concerned with establishing the LAN connection, transferring data, and terminating the connection; this layer is nonhardware-specific.

management plane (M plane)   The part of the ATM layered architecture that provides control of an ATM node and consists of plane management and layer management.

media access unit (MAU)   A wiring concentrator or hub.

medium access control layer (MAC)   The hardware-specific protocol layer within a network interface card's chip that's concerned with transmitting data across a LAN.

mesh topology   A network topology in which each node is linked to several other nodes, so there's no single point of failure.

minimum cell rate (MCR)   The rate (cells/second) of an application's ability to handle latency.

multiprotocol access services (MAS)   The protocol support for IBM's switched virtual networking (SVN).

multiprotocol over ATM (MPOA)   An architecture being developed that will support network-layer switching at all levels of an ATM-based network so products can be developed to act as virtual distributed routers.

network interface card   An adapter card that contains the IEEE 802.2 protocols for the data link layer in the form of a chip.

network layer   The OSI model layer responsible for network routing, correct sequencing of packets, and implementing quality-of-service requirements requested by the transport layer.

network-to-network/network-to-node (NNI) specification   This ATM Forum specification defines the interface between two switches or between two networks.

node   An intelligent device, generally a computer, on a network.

noncontention network   A network such as Token Ring where nodes don't compete for network access.

open shortest path first (OSPF) protocol   A routing protocol designed for TCP/IP that overcomes many of the limitations of the routing information protocol (RIP), including the way it handles broadcasts and its ability to handle more than 15 routers in a path.

open systems interconnect (OSI) model   A model consisting of layered protocols to facilitate communications between computer networks.

packet assembler/disassembler (PAD)   A device that provides protocol translation for data so it can travel over a packet-switched network such as an X.25 wide area network.

packet-level discard   A FORE Systems switch feature that results in discarding groups of cells that form entire packets rather than discarding random cells from multiple packets.

partial packet discard (PPD)   A FORE Systems feature that discards all remaining cells in the packet (the tail of the packet) when there's a buffer overflow and a cell must be discarded.

payload type identification (PTI)   The three-bit ATM field used to identify both the payload type carried in a cell and control procedures.

peak cell rate (PCR)   The maximum ATM data rate that a connection will support without losing data.

permanent virtual circuit (PVC)   An ATM circuit that must be set up in advance, unlike SVCs, which are established dynamically.

permanent virtual path (PVP) tunneling   A Cisco software feature that enables a network manager to set up temporary switched virtual connections over previously established PVPs.

physical layer   The OSI model layer responsible for the physical transmission of data. It's also a layer in ATM that's responsible for transmitting and receiving data, grouping cells in payload envelopes, and adding routing information.

physical-medium-dependent (PMD) sublayer   The physical medium portion of ATM's physical layer that's responsible for supporting different physical media, such as fiber or twisted-pair wire.

point of presence (POP)   Interface points to long-distance carriers found within a LATA.

presentation layer   The OSI model layer concerned with how information is physically displayed or presented.

primary rate interface (PRI)   Multiple access into an ISDN network that includes 23 bearer channels of 64 Kbps each and one D channel of 64 Kbps.

quality of service   Quality of ATM service based on the negotiation of such variables as cell loss ratio, cell delay, and delay variance.

quantum flow control (QFC)   A credit-based approach for ATM traffic management supported by Digital Equipment Corporation.

resource reservation protocol (RSVP)   A Cisco-supported internetwork protocol that allows an application to dynamically reserve resources for different classes of service.

router   A network device (hardware or software) that routes packets based on the network layer protocols of the OSI model.

routing information protocol (RIP)   A routing protocol used by Berkeley-derived UNIX systems.

rules-based VLANs   Virtual LANs created on the basis of rules established by the network manager.

session layer   The OSI model layer concerned with the mode of transmission and synchronizing communications.

smart buffers   Buffers that manage bandwidth and efficiently prioritize traffic into multiple service levels.

source routing   When a node solicits information that helps it build a roadmap to the data destination where it wants to route data.

store-and-forward switching   A method in which address tables are used to keep track of devices linked to a switch's ports; the entire packet is read before it's forwarded in an error-free manner.

sustained cell rate (SCR)   The average ATM cell throughput rate an application is permitted.

switched multimegabit data service (SMDS)   A connectionless service based on the IEEE 802.6 standard designed for metropolitan area networks running applications that generate frequent but short data streams.

switched virtual circuit (SVC)   An ATM circuit that can be established dynamically.

switched virtual networking (SVN)   A comprehensive IBM model for building and managing switch-based multiprotocol networks, including ATM networks.

switching hub   An intelligent hub with an installed switching module that allows data to be switched among LAN segments within the hub.

synchronous optical network (SONET)   A standard to synchronize public communications networks and tie them together via a high-speed fiber-optic links.

T-1   A trunk that contains 24 channels, each channel with a capacity of 64,000 bps, and one 64,000-bps channel for error checking—for an aggregate bandwidth of 1.544 Mbps.

T-3   A trunk that contains 28 T-1 trunks for an aggregate bandwidth of 45 Mbps.

tagging   A method for switches to update each other in which a tag containing a node's virtual LAN identifier is attached to Ethernet frames.

Telecommunications Standardization Sector (ITU-T)   Formally known as the CCITT, this UN treaty organization focused on international specifications for B-ISDN/ATM, which is now the basis of ATM.

time-division multiplexer (TDM)   A multiplexer that guarantees a data time slot for each of the channels it multiplexes.

transmission convergence (TC) sublayer   That portion of the ATM physical layer associated with the transmission and reception of cells, including cell delineation, cell scrambling/descrambling, and cell-rate decoupling.

transparent bridging   The type of bridge that checks addresses and then forwards packets meant for a different segment.

transport layer   The OSI model layer responsible for providing the quality of service requested by the network layer.

unspecified bit rate (UBR)   The ATM protocol that defines traffic with no specified bit rate and no quality-of-service guarantees.

user plane (U plane)   The part of ATM's layered architecture used for end-to-end or user-to-user data transfer.

user-to-network interface (UNI)   This ATM Forum specification defines the interface between users of ATM services and ATM network nodes.

variable bit rate (VBR)   Designed for bursty traffic, this ATM protocol regulates service for variable bit-rate traffic; both a peak cell rate and a sustained cell rate must be negotiated.

virtual channel identifier (VCI)   This value identifies a single virtual channel on a particular virtual path.

virtual LAN (VLAN)   A LAN that's organized logically rather than by nodes' physical locations.

virtual network (VNET)   Digital Equipment Corporation's definition of a collection of VLANs interconnected with high-performance routing.

virtual path (VP)   The path selected for the unidirectional transport of ATM cells belonging to the virtual channels associated with a common identifier value.

virtual path identifier (VPI)   A value that appears in an ATM cell header that identifies a bundle of one or more virtual channels.

virtual segment VLAN   A virtual LAN created on the basis of MAC addresses that in effect is a single subnet functioning at layer 3 of the OSI model.

virtual subnet VLAN   A virtual LAN that's created on the basis of layer-3 subnet addresses, so nodes can be segmented based on such network layer characteristics as protocol.

weighted fair queuing   A feature that enables a network to identify various types of network traffic and assign them consistent routing priorities that are weighted to minimize latency.

wide area network (WAN)   A network that links together networks located in different geographical areas.

workgroup LANs   LANs that link together groups of users that share a common application, such as CAD/CAM/CAE.

Index

ABOUT THE AUTHOR

Stan Schatt (La Jolla, California) is the author of *Linking LANs* and *Understanding Network Management*, both for McGraw-Hill, as well as over 20 other books, including the bestselling *Understanding Local Area Networks*, now in its fourth edition. As LAN/WAN Research Director for Computer Intelligence InfoCorp., Schatt's primary responsibility is to track the LAN market and provide accurate analysis and forecasts for major LAN vendors. He is widely quoted in all the major trade publications, including *PC Week*, *Network World*, *Computerworld*, and *LAN Times*. Schatt holds a Ph.D. from USC and is former chairman of the Telecommunications Management Department of DeVry Institute of Technology.